FAST CARS, CLEAN BODIES

OCTOBER BOOKS

Annette Michelson, Rosalind Krauss, Yve-Alain Bois, Benjamin Buchloh, Hal Foster, Denis Hollier, and John Rajchman, editors

Broodthaers, edited by Benjamin H. D. Buchloh

AIDS: Cultural Analysis/Cultural Activism, edited by Douglas Crimp

Aberrations, by Jurgis Baltrušaitis

Against Architecture: The Writings of Georges Bataille, by Denis Hollier

Painting as Model, by Yve-Alain Bois

The Destruction of Tilted Arc: *Documents,* edited by Clara Weyergraf-Serra and Martha Buskirk

The Woman in Question, edited by Parveen Adams and Elizabeth Cowie

Techniques of the Observer: On Vision and Modernity in the Nineteenth Century, by Jonathan Crary

The Subjectivity Effect in Western Literary Tradition: Essays toward the Release of Shakespeare's Will, by Joel Fineman

Looking Awry: An Introduction to Jacques Lacan through Popular Culture, by Slavoj Žižek

Cinema, Censorship, and the State: The Writings of Nagisa Oshima, by Nagisa Oshima

The Optical Unconscious, by Rosalind E. Krauss

Gesture and Speech, by André Leroi-Gourhan

Compulsive Beauty, by Hal Foster

Continuous Project Altered Daily: The Writings of Robert Morris, by Robert Morris

Read My Desire: Lacan against the Historicists, by Joan Copjec

Fast Cars, Clean Bodies: Decolonization and the Reordering of French Culture, by Kristin Ross

FAST CARS, CLEAN BODIES

DECOLONIZATION AND THE
REORDERING OF FRENCH CULTURE

KRISTIN ROSS

AN OCTOBER BOOK

THE MIT PRESS
CAMBRIDGE, MASSACHUSETTS
LONDON, ENGLAND

First MIT Press paperback edition, 1996

This book was set in Bembo by DEKR Corporation and was printed and bound in the
United States of America.

Library of Congress Cataloging-in-Publication Data

Ross, Kristin.
   Fast cars, clean bodies : decolonization and the reordering of French culture / by
Kristin Ross.
      p.   cm.
   "An October book."
   Includes bibliographical references and index.
   ISBN 0-262-18161-4 (HB), 0-262-68091-2 (PB)
      1. France—Civilization—1945– 2. Decolonization—France—History—20th
century. 3. Technology—Social aspects—France—History—20th century.
4. Popular culture—France—History—20th century. 5. France—Civilization—
American influences. 6. Racism—France. 7. French literature—20th century—
History and criticism. 8. Authors, French—20th century—Political and social views.
I. Title.
DC33.7.R65   1995
944.083—dc20   94-17815
                                                                              CIP

10 9 8 7 6 5

In memory of Roberto Crespi

Contents

CONTENTS

ACKNOWLEDGMENTS

My thinking about the culture of postwar French modernization dates back to an issue of *Yale French Studies* that Alice Kaplan and I edited on "everyday life"; I am grateful for her support at an early stage of this project, as well as that of Fredric Jameson, Franco Moretti, and Ann Smock. Thanks also to Denis Echard, Denis Hollier, Linda Orr, Danielle Rancière, Jacques Rancière, Adrian Rifkin, and the UCSC graduate students in the fall 1992 "strike" seminar, for tips along the way.

Research for this book was conducted in France with the aid of grants from the American Council of Learned Societies and from the University of California President's Foundation; and in California with the help of the *fabulous* interlibrary loan staff at UC Santa Cruz. I am grateful to Jenya Weinreb at The MIT Press for her help in preparing the manuscript. An earlier version of chapter 2 appeared in *October* 67 (Winter 1994).

Thanks to Cindi Katz and Neil Smith at Rutgers, and to Leo Bersani and Ann Smock at Berkeley, for inviting me to try out an early version of my material about automobiles, and to the Cultural Studies Center at UC Santa Cruz for providing another happy forum for work in progress.

Page duBois, Alice Kaplan, and Carter Wilson read the manuscript in its entirety, offering new perspectives, encouragement, and debate.

Finally, my deepest thanks to Harry Harootunian for his unflaggingly enthusiastic yet critical readings of this book in all of its many stages.

FAST CARS, CLEAN BODIES

In Claude Chabrol's second movie, *Les cousins* (1959), a young provincial boy called Charles arrives in Paris to study law, and shares an apartment with his cynical, worldly, "Nietzschean" cousin, also a law student. While his debauched cousin pursues a frenetic social life, the country boy spends most of his time in his room writing fond, descriptive letters back to his mother in the village; tiring momentarily of this, he decides to read some Balzac. The bookstore owner is so pleased with his choice ("all the rest of them, they just want to read pornography and detective fiction") that he makes him a present of a copy of *Illusions perdues*.

Françoise Giroud, one of the key figures behind the proliferation of women's magazines in the 1950s in France and an important character in this book, recalls in her memoirs how she and the cofounder of *Elle* magazine, Hélène Lazareff, imagined the ideal reader of their new magazine as the first issue hit the stands. The reader envisioned by the staff at *Elle* was most likely young, between twenty-five and thirty-five, tired of wartime deprivation, in need of frivolity, and she lived in Angoulême. Why Angoulême? I don't remember, says Giroud. Perhaps because of Rastignac.[1]

In a series of articles that later came to be read as the manifesto of the French novel of the late 1950s, Alain Robbe-Grillet situates his own era and its realist mode of representation by comparing it with that of Balzac. Balzac's period was marked by "the apogee of the individual," whereas today is the period of "administrative numbers." The objects that appear in Balzac's descriptions stagger under the weight of all that they are meant to signify; Robbe-Grillet's objects are present in and for themselves, unencumbered by human significance. Balzac, for Robbe-Grillet, represents "the old myths of depth"; Robbe-Grillet proposes instead "a flat and discontinuous universe where each thing refers only to itself."[2]

Yet if Balzac and his mode of narrative representation provide Robbe-Grillet with the example of everything that the novels of the day should not now be, still Balzac's claim to have represented his own era accurately, realistically, and with authority goes unquestioned. In fact, Robbe-Grillet yearns to *be* the Balzac of his day, to follow his example and produce a new, modernized mode of realism suitable to representing the "new man" and his era of numbers. The New Novel would be a Human Comedy without the humans.

The Balzac of his day, the Rastignac of her day, the Lucien de Rubempré of the present. In the late 1950s and early 1960s—the roughly ten-year period I examine in this book, the years after electricity but before electronics—Balzac provides a way for people to establish the particular hopes, anxieties, fears, and aspirations of their own era; he is a recurrent figure in an allegory by way of which the present appears as both a repetition and a difference, a means of continuity and a mark of rupture. Once more, as in the 1850s, the countryside is being depleted, and villagers flock to the new forms of employment, opportunities, and pleasures that can be found in the cities. But the newly arrived Parisians of the postwar era are likely to be provincial French women come to work as shopgirls as in Chabrol's *Les bonnes femmes* (1960) or in Elsa Triolet's *Roses à crédit* (1959); village boys such as Charles who come to take an advanced degree at the moment when higher education is no longer the prerogative of a tiny elite; or Algerian immigrants seeking work in the car factories on the outskirts of Paris as in Claire Etcherelli's *Elise ou la vraie vie* (1966). Other realist characters have changed as well. The furtive calculations and the limited horizons of the Balzacian "type" par excellence, the notary, are both repeated and surpassed by another kind of supreme calculator—one who by the very development of his discipline becomes an autonomous factor of the postwar acceleration: the engineer. "And so beyond the engineer whose knowledge increases and whose machines perfect themselves and multiply, a manner of looking at things is forming, and soon a whole way of reasoning that marks our era."[3] The stable old, propertied *honnête bourgeois* of Balzac's era reappears in a very different, streamlined, and fast-moving format: the forward-looking, hard-working *jeune cadre*. And yet in *Les belles images* (1966) Simone de Beauvoir will uncover the strands of class interest that unite the two, reveal them to be the same man wearing different masks.

The essence of the recurring Balzac allegory in the decade I study in this book has to do with periodization. As formulated by Alain Touraine, it is an argument that presumes the epochal originality of Balzac's time in order to argue the same status for the present: "At the dawning of French industrialization, Balzac was aware of the frenzy for money, the social upheaval, but 1848 had to arrive before all the problems surrounding industrial work and the proletariat could be seen in the light of day. Aren't we now, within the new society being organized before our eyes, existing in a moment comparable to the one in which Balzac wrote?"[4] Following Touraine's analogy, May '68 would be the new 1848, the confirming afterthought, the event that certified the massive social upheaval and land grab of the decade that preceeded it. With the largest strikes in French history, May '68 would bring all the problems and dissatisfactions surrounding the French lurch into modernization to the light of day. It was the event that marked the political end of that accelerated transition into Fordism: a protest against the Fordist hierarchies of the factories and the exaggerated statism that had controlled French modernization. (The economic confirmation of the end would come a few years later with the oil crisis and economic recession of the early 1970s.)

If I have stopped short of a consideration of the events of May '68 in this book, it is because I wanted to consider instead the event of French modernization in the decade that came before—to consider, that is, French modernization *as* an event. Modernization is, of course, not an event but a process, made up of slow- and fast-moving economic and social cycles. But in France the state-led modernization drive was extraordinarily concerted, and the desire for a new way of living after the war widespread. The unusual swiftness of French postwar modernization seemed to partake of the qualities of what Braudel has designated as the temporality of the event: it was headlong, dramatic, and breathless. The speed with which French society was transformed after the war from a rural, empire-oriented, Catholic country into a fully industrialized, decolonized, and urban one meant that the things modernization needed—educated middle managers, for instance, or affordable automobiles and other "mature" consumer durables, or a set of social sciences that followed scientific, functionalist models, or a work force of ex-colonial laborers—burst onto a society that still cherished prewar outlooks with all of the force, excitement, disruption, and horror of the genuinely new.

It is this swiftness that fascinated me, and that I recall being made aware of when I first read Henri Lefebvre. Contrasting the French experience to the slow, steady, "rational" modernization of American society that transpired throughout the twentieth century, Lefebvre evoked the almost cargo-cult-like, sudden descent of large appliances into war-torn French households and streets in the wake of the Marshall Plan. Before the war, it seemed, no one had a refrigerator; after the war, it seemed, everyone did. Fordist consumption, as Michel Aglietta points out (and as the organization of this book reflects) is governed by two commodities: "the *standardized housing* that is the privileged site of individual consumption; and the *automobile* as the means of transport compatible with the separation of home and workplace."[5] French people, peasants and intellectuals alike, tended to describe the changes in their lives in terms of the abrupt transformations in home and transport: the coming of objects—large-scale consumer durables, cars and refrigerators—into their streets and homes, into their workplaces and their *emplois du temps*. In the space of just ten years a rural woman might live the acquisition of electricity, running water, a stove, a refrigerator, a washing machine, a sense of interior space as distinct from exterior space, a car, a television, and the various liberations and oppressions associated with each. What were the effects of such a sudden series of changes? Where were these effects best registered, recorded? Who bore the costs? Modern social relations are of course always mediated by objects; but in the case of the French, this mediation seemed to have increased exponentially, abruptly, and over a very brief period of time. If I return throughout the book to the films of Jacques Tati, it is because they make palpable a daily life that increasingly appeared to unfold in a space where objects tended to dictate to people their gestures and movements— gestures that had not yet congealed into any degree of rote familiarity, and that for the most part had to be learned from watching American films. Was it a mark of the particular rapidity of French modernization that so much of the country's intellectual effort of the period—the earliest (and thus most materialist) works by Barthes and Baudrillard, for example, or that of the Situationists, Cornelius Castoriadis, Edgar Morin, or Maurice Blanchot in his review essays of Lefebvre—took the form of a theoretical reflection on "everyday life"? Or that "everyday life" is elevated to the status of a theoretical concept only at this particular conjuncture? Theoretical categories are not free-floating analytic devices, innocent of historical content. If they instead find their origins in forms of experience, then the transitory importance of critical categories like "alienation" and "everyday life," or the move to the forefront of the concept of "reification" during these years, must then be another sign of the upheaval in social relations occasioned by the sudden, full-scale entry of capital into "style of life," into lived, daily, almost imperceptible rhythms.[6] This is no less true for the dominant conceptual apparatus as well. A key ideological concept like "communication," for example, began to refer in mid-century not only to the dawning of the new information technologies but to the ideal spatial arrangement of rooms in modern suburban homes; it was also the title of the leading journal of the day devoted to advances in structuralism. The word *communication* was everywhere—and yet the experience of communication itself, be it understood as spontaneous expression, reciprocity, or the contiguity necessary for reciprocity to exist, was precisely what was in the process of disappearing under the onslaught of merchandise and the new forms of media technologies. Merchandise (or exchange relations) is first of all the production of nonexchange between people; structuralism, the dominant intellectual movement of the period, fetishizes "communication" at the very moment when various forms of direct, unmediated relations (*communicare,* Latin: to be in relation with) among people are waning or being decisively transformed.

Touraine's analogy then holds; his era is the dawning of a new economic and social era in France comparable to that of the beginnings of French industrialization in the 1830s and 1840s. Economists agree that the consolidation of a Fordist regime in France in the decade or so before 1968—a period of "growth without precedent of capitalism in France,"[7] the peak decade, that is, of the thirty-year postwar economic boom—was an extraordinarily voluntarist and thus wrenching experience. It took place, for instance, at the cost of a relentless dismantling of earlier spatial arrangements, particularly in Paris where the city underwent demolitions and renovations equivalent in scale to those Haussman oversaw a hundred years earlier.[8] And it transpired in the decade that saw the stumbling and final collapse of the French Empire, from the decisive battle of Dien Bien Phu in the spring of 1954, to the first major Algerian uprisings a few months later, to the referendum on African independence in 1958, to the granting of that independence in 1960, all the way through to the Evian Accords that officially announced the hard-won independence of Algeria in May 1962.

Touraine makes no mention of the end of the empire in his characterization of the singularity of the age (nor does he mention the beginning of the empire in reference to Balzac's). His omission is characteristic of many such narratives of the period that tend, even today, to choose between the two stories, the story of French modernization and Americanization on the one hand, or the story of decolonization on the other.[9] I have tried instead throughout this book to hold the two stories in the tension of what I take to be their intricate relationship as it was lived then and as it continues into the present. The peculiar contradictions of France in that period can be seized only if they are seen as those of an exploiter/exploited country, dominator/dominated, exploiting colonial populations at the same time that it is dominated by, or more precisely, entering more and more into collaboration or fusion with, American capitalism. It is this particular tension, in fact, as I argue in the final section of the book, that makes the emergence of the character and social type of the *jeune cadre,* that high priest of Fordism, something of a national allegory for the modernizing France of the 1950s and 1960s. Midway between owner and worker, managing the proletariat but punching a time clock too, the *cadre,* like France itself, was a "dominated agent of capitalist domination."[10]

Thinking the two narratives together means taking seriously the catchphrase popularized by Lefebvre and the Situationists in the early 1960s: "the colonization of everyday life." In the case of France, in other words, it means considering the various ways in which the practice of colonialism outlived its history. With the waning of its empire, France turned to a form of interior colonialism; rational administrative techniques developed in the colonies were brought home and put to use side by side with new technological innovations such as advertising in reordering metropolitan, domestic society, the "everyday life" of its citizens.

Marxist theory had made considerable progress in refining theories of imperialism in the domain of international relations. Lefebvre now pushed that theory to apply the insights garnered from an international analysis to new objects: to the domain of interregional relations within France, for example, or the space of domesticity and practices of consumption. But it was above all the unevenness of the built environment of the city, its surroundings, and its social geography that came to crystallize, for Lefebvre, the contradictions of postwar life. For speculative capital, no longer drawn to foraging abroad, was increasingly directed toward investment in the built environment: Paris, the city itself, became the new site for a generalized exploitation of the daily life of its inhabitants through the management of space. At times the conversion from exterior to interior colonialism was facilitated by a literal transfer of personnel; thus, a city councilor involved in the Parisian renovation debates of the early 1960s remarked, "France decolonized the Third World while colonizing Paris, appointing as head of the commission charged with making decisions about the capital functionaries who had made their careers in Black Africa or in Asia."[11] But such literal transfers in personnel pale in importance when compared to the emergence, in those years, of what might be termed a *comprador* class serving the interests of the state: financiers, developers, speculators, and high administrative functionaries. Modernization brought into being a whole new range of middlemen and go-betweens, new social types that dominated and profited from the transformations wrought by the state. The *jeune cadre* elevated to an intermediate position in the corporate hierarchy, the housewife elevated to the role of technician or manager of the newly modern home—couldn't these social ascendancies, too, be seen in the light of a generalized "compradorization" of the French middle class?

In the France of today the tendency to "keep the two stories separate" has, I think, very serious social and political consequences, consequences that are being played out in the rise of the various neoracisms of the 1980s and 1990s that focus on the figure of the immigrant worker.

Keeping the two stories apart is usually another name for forgetting one of the stories or for relegating it to a different time frame. This is in fact what has occurred. For, from this perspective (a prevalent one in France today), France's colonial history was nothing more than an "exterior" experience that somehow came to an abrupt end, cleanly, in 1962.[12] France then careered forward to new frontiers, modern autoroutes, the EEC, and all-electric kitchens. Having decisively slammed shut the door to the Algerian episode, colonialism itself was made to seem like a dusty archaism, as though it had not transpired in the twentieth century and in the personal histories of many people living today, as though it played only a tiny role in France's national history, and no role at all in its modern identity. One of the arguments of this book is that the very logic of (racial) exclusion that would "keep the two stories separate" is itself the outcome of the accelerated capitalist modernization the French state undertook in those years. The new contemporary racism centering on questions of immigration is, as the contemporary detective stories of Didier Daeninckx make clear, a racism that has its roots in the era of decolonization and modernization, in the inversion of movements of population between the old colonies and the old metropoles, in the conflict that crystallized in those days between the modern and the unmodern (or traditional)—the latter being directly referred to race and supposedly racial traits, such as laziness or filth. The immigration that haunts the collective fantasies of the French today is the old accomplice to the accelerated growth of French society in the 1950s and 1960s. Without the labor of its ex-colonial immigrants, France could not have successfully "Americanized," nor competed in the postwar industrial contest. In the economic boom years, in other words, France made use of the colonies "one last time" in order to resurrect and maintain its national superiority over them—a superiority made all the more urgent by the ex-colonies' own newly acquired nationhood.

If the colonies provided the labor, the fuel came from the West. Immediately after the war a particular fantasy was exported by the

United States, along with the gadgets, techniques, and experts of American capitalism, to a Europe devastated by war: the fantasy of timeless, even, and limitless development. Capitalist modernization presents itself as timeless because it dissolves beginning and end, in the historical sense, into an ongoing, naturalized process, one whose uninterrupted rhythm is provided by a regular and unchanging social world devoid of class conflict. In this book I show how the arrival of the new consumer durables into French life—the repetitive, daily practices and new mediations they brought into being—helped create a break with the eventfulness of the past, or better, helped situate the temporality of the event itself as a thing of the past. I have also argued the complicity of much of the French intellectual production of the era—from structuralism to the Annales school of historiography—with that dissolution, because of the way in which these sciences eliminate from their horizon everything that might conceivably upset the processes of repetition, the way in which they have abandoned the event as a conceptual category. My own somewhat perverse consideration of French modernization as an event is an attempt to fly in the face of this still hegemonic practice. By historicizing France's transition into American-style mass culture, the prehistory of its postmodernism, I try to provide an experience of the historicity that theories of postmodernism, themselves rooted in the intellectual developments of the 1950s and 1960s and in the dissolution of the event and of diachronic agency, seek to efface.

But we must return now briefly to the most important promise made by modernization: its evenness. Modernization is even because it holds within itself a theory of spatial and temporal convergence: all societies will come to look like us, all will arrive eventually at the same stage or level, all the possibilities of the future are being lived now, at least for the West: there they are, arrayed before us, a changeless world functioning smoothly under the sign of technique. The process of development in the West has been completed; what comes now is already in existence: the confused syncretism of all styles, futures, and possibilities. Modernization promises a perfect reconciliation of past and future in an endless present, a world where all sedimentation of social experience has been leveled or smoothed away, where poverty has been reabsorbed, and, most important, a world where class conflict is a thing of the past, the stains of contradiction washed out in a superhuman hygienic effort, by new levels of abundance and equitable distribution.

And yet the French experience, in its highly concentrated, almost laboratory-like intensity, has the advantage of showing modernization to be instead a *means* of social, and particularly racial, differentiation; a differentiation that has its roots in the 1950s discourse on hygiene I examine in the second chapter of this book and take up again in the third. If the consolidation of a broad middle class more or less transpires during these years, it is also during these years that France distances itself from its (former) colonies, both within and without: this is the moment of the great cordonning off of the immigrants, their removal to the suburbs in a massive reworking of the social boundaries of Paris and the other large French cities. On the national level France retreats within the hexagon, withdraws from empire, retrenches within its borders at the same time that those boundaries are becoming newly permeable to a whirlwind of economic forces—forces far more destructive of some received notion of "national culture" than any immigrant community could muster. The movement inward—a whole complex process that is in some ways the subject of each of my chapters and that Castoriadis, Morin, and Lefebvre all called "privatization"—is a movement echoed on the level of everyday life by the withdrawal of the new middle classes to their newly comfortable domestic interiors, to the electric kitchens, to the enclosure of private automobiles, to the interior of a new vision of conjugality and an ideology of happiness built around the new unit of middle-class consumption, the couple, and to depoliticization as a response to the increase in bureaucratic control of daily life. Modernization requires the creation of such a privatized and depoliticized broad middle strata: a "national middle class"; from this point on, national subjectivity begins to take the place of class. Now, in our own day, when the broad middle strata has become coterminous with the nation itself in France, more atavistic logics or principles of exclusion are coming to light. Class conflict, after all, implies some degree of negotiability; once modernization has run its course, then one is, quite simply, either French or not, modern or not: exclusion becomes racial or national in nature. If the ideology of modernization says convergence—all societies will look the same—what it in fact sustains and freezes into place is the very unevenness or inequality that it was supposed to overcome: they will never be like us, they will never catch up. In today's Paris that frozen temporal lag appears as a spatial configuration: the white, upper-class city *intra muros,* surrounded by islands of immigrant communities a long RER ride away.

Touraine's analogy falters when we look to find the writer who foresaw and undertook the monumental task of representing such a momentous transition. Despite his ambitions, Robbe-Grillet did not turn out to be the Balzac of his day. And Didier Daeninckx, who offers the most acute contemporary arguments that the conditions of the immediate present lie in the failures and events of the 1950s and 1960s, is a writer of today, not then. Perhaps the point is that no single writer could occupy the position of Balzac in a moment that was also characterized by the introduction of market research into book publishing, by the mass-marketing of paperback books, by the dawning of image culture, and by a profound crisis in the traditional novel that itself reflected the new fragmentation of social life. But Robbe-Grillet's novels and theoretical reflections, in particular, are themselves too imbued with the ideology of modernization to offer the necessary critical perspective; as Jacques Leenhardt's work has shown, the New Novel is part and parcel of that ideology, and of the whole contemporary movement whereby a naive or vulgar materialism comes to be substituted for dialectical materialism, and *mentalité* (or shared culture, shared values, or any of a number of prevalent designations of "consensus" or averaging) takes the place of ideology. Like structuralism and the Annales school of historiography, the New Novel is complicitous with the workings of capitalist modernization, in part because of its avant-gardist refusal or dismantling of historical narrative.

For help in formulating a critical prehistory of postmodernism in France I have had to look elsewhere: to those artists and thinkers who historicized their era at the time and who gave full voice to the debates and controversies surrounding modernization. Novelists such as Christiane Rochefort, Simone de Beauvoir, and Georges Perec working in a realist mode; filmmakers from Jacques Tati to Jacques Demy; and those social theorists who turned their attention after the war to "everyday life" performed the labor of accounting for the present—its disruptions and its social costs—that the historians, lost in a prolonged dream about the *longue durée* of feudalism, chose to avoid. If the single monumental realist author working to represent the totality of an era—a "Balzac"—has been relegated to a definitive past, then it is still to the realist *mode* that we must look to find the narrative style best suited to portraying unevenness.[13] The realist mode attempts to come to terms with, or to give an historical account of, the fatigue and exhilaration of moments when people find themselves living two lives at once. As Raymond Williams has suggested, realism gives a shape to the experiences of those on the outer edges of modernization's scope, the ones caught just outside or the ones who have been left behind, the ones for whom abundance is accompanied by a degradation in their conditions of existence. Realism offers a voice to those who live in a different temporality, who follow a pace of life that is nonsynchronous with the dominant one. In the postwar period realist fiction and film offered a critique of official representations of a uniformly prosperous France, surging forward into American-style patterns of consumption and mass culture. It is in these works that we can still glimpse the "democracy of consumption" for what it is: the newest form of bourgeois democracy, the alibi of a class society.

LA BELLE AMÉRICAINE

CAR HISTORY

Not only was 1992 the year of the grand opening of Eurodisney in France; on March 27th of the same year, the assembly lines of the Renault factory at Billancourt on the southwest outskirts of Paris ground to a halt after ninety years of activity. What was for over half a century the largest factory in the most powerful French industry—the Régie Renault—had in recent years been reduced to a mere annex in a now decentralized and successfully internationalized enterprise: in 1992 Renault negotiated a lucrative alliance with the Swedish firm of Volvo.

The Renault firm, which began at the turn of the century as an *atelier* in a family garden, was by World War I already the second largest automobile builder in France. Of the three brothers who launched the family business, he who became known as Louis "Vivre c'est consommer" Renault quickly emerged as its creative head. World War I, during which Billancourt was transformed into an armament factory, producing 55 percent of the tanks used during the war, propelled his enterprise forward to the level of industry giant. To expand his work site, Renault began to buy up all of the houses and available surfaces on the Ile Seguin, and by 1929 the once-bucolic garden-covered island immortalized in Corot landscapes and best known as a destination for weekend boat trips from Paris had become "the factory of tomorrow," what one journalist described as "a mountain of cement [where one hears] the rumbling of a thousand machines, where men make the same precise movement eight hours a day all along a line and enormous chimneys spew forth smoke."[1] Over the years Louis Renault erected guard walls to protect his island factory from the river waves, and gradually Billancourt took on its modern appearance as a powerful fortress in the Seine.

The concentration of over 30,000 workers at Billancourt made it the privileged arena for the rise of union militancy in France; the presence of such a massive concentration of workers just a few minutes from Paris turned the factory into the ideal "laboratory" for industry analysts, bosses, intellectuals, and the press to observe the transformations in professional, syndical, and political life occurring in France. In 1913, for example, the importation of a very partial Taylorism from the United States—*chronométrage*—caused the first savage strikes, one CGT leader denouncing the system as "the organization of exhaustion";[2] Louis Renault became legendary for his severe methods of surveillance, and in the 1930's he would frequently order the entire factory evacuated by police.[3] With the Liberation and the nationalization of the Renault factory in 1945, the marriage between the symbol of the car industry and the symbol of the worker was complete: throughout the thirty-year postwar industrial boom that followed—*les trente glorieuses*—the *métallo* or car worker at Billancourt came to serve as the incarnation of the working class. The CGT (*Confédération générale du travail* or confederation of French unions) and the Parti Communiste spoke in the worker's name, and, in return, any worker from Billancourt was listened to attentively and tended to rise in the party ranks. In 1945 half of the Billancourt workers were in the CGT.

Both the automobile industry and the intellectual industry were centered in Paris, making Billancourt the site for a number of interesting meetings; throughout the century it was the place intellectuals went to encounter "the people." We all remember the crucial role played by Renault in May '68, and have probably seen any of a number of famous photos of Sartre at the factory gate. The familiar slogan of solidarity intoned throughout the 1960s and 1970s, "Il ne faut pas désespérer Billancourt," where Billancourt stands in as metonymy for the entire working class, originated as a line spoken in one of Sartre's plays. But Billancourt was also the location for Simone Weil's investigation into industrial working conditions in the 1930s, an investigation that produced the first of what would become a long line of "assembly-line testimonios," her 1934 *La condition ouvrière;*[4] Billancourt was also the subject of perhaps the most stunning French photography produced in the last few decades, René-Jacques's documentation of the birth of the 4CV in the early 1950s.[5] But the factory was probably most at the forefront of public attention during the Maoist experiments of "l'établissement" in the early 1970s, the moment when Sorbonne professors got jobs on the line in order to organize from the inside, when Simone Signoret paid supportive visits to hunger strikers, when Jean-Luc Godard, Yves Montand, and Jane Fonda filmed *Tout va bien* based on "All is well" the sabotage activities of workers on the line. The turning point of the Maoist movement was undoubtedly the death, in March of 1972, of fired Renault worker Pierre Overney, who was killed by a security guard at the factory gate; in the wake of his death—some 200,000 people joined in his funeral procession through the streets of Paris—a Maoist commando group seized and held in an undisclosed location the chief personnel officer at Billancourt, releasing him unharmed two days later. That labor unions should denounce this operation was to be expected; however, when other far left groups joined in on the denunciation, it became virtually impossible to do political work at Billancourt after the affair subsided.[6] The death of Pierre Overney was traumatic—in and of itself and also because of the absence of reaction among the workers at Renault. It marked the end of a certain kind of hope in the possibility

FIGURE 1.1 René-Jacques, Usines Renault, Billancourt, 1951. Courtesy Ministère de la Culture—France.

of constituting a new revolutionary worker force in the factories; many would say it marked the end of the 1960s in France.

At a time when it is fashionable to leap beyond the materiality of an object to enter into the mirror of symbolic relations, I want to stay fairly close, in this chapter, to the central vehicle of all twentieth-century modernization, the automobile. In the middle of this century, the automobile industry, more than any other, becomes exemplary and indicative; its presence or absence in a national economy tells us the level and power of that economy. Postwar French economic growth was a direct result of having modernized sectors of production that were seen to be vital—and the most vital of these was automobile production. In turn, the augmentation in French buying power after 1949 was used principally to buy cars. But to recount these simple facts is already to argue that the car *is* the commodity form as such in the twentieth century, an argument that becomes all the more convincing when we remember that "Taylorization"—the assembly line, vertical integration of production, the interchangeability of workers, the standardization of tools and materials—"Taylorization" was developed *in the process of producing* the "car for the masses" and not the inverse.

The very fact of its being the commodity form as such tends to consign the car to the edges of historical discourse—despite the now common use of terms from the history of its production, "Fordism," and "post-Fordism," to designate a kind of twentieth-century periodization. I think this has less to do with the ubiquity or banality of the object—its now seamless integration into the fabric of everyday life—than with the way in which historicity is, so to speak, "emptied out" at each of the three "moments"—production, transformation into discourse (i.e., advertising, media representations), and consumption and use—that define the car. For the car is not only implicated in a certain type of mobilization by capital, it is also an active though partial agent in the *reproduction* of that structure—thus its embeddedness, in each of the three

19

"moments," in a temporality of repetition. In the first of these, production, repetition takes the form of the inexorable movement of the assembly line itself, whose synchronic systematicity and precise reenactment of the same creates in the line worker an experience of timelessness, of the same story told again and again. The sensation is best described by Robert Linhart:

> The assembly line isn't as I'd imagined it. I'd visualized a series of clear-cut stops and starts in front of each work position: with each car moving a few yards, stopping, the worker doing his job, the car starting again, another one stopping, the same operation being carried out again, etc. I saw the whole thing taking place rapidly—with those "diabolical rhythms" mentioned in the leaflets. The assembly line: the words themselves conjured up a jerky, rapid flow of movement.
>
> The first impression, on the contrary, is one of a slow but continuous movement by all the cars. The operations themselves seem to be carried out with a kind of resigned monotony, but without the speed I expected. It's like a long, gray-green gliding movement, and after a time it gives off a feeling of somnolence, interrupted by sounds, bumps, flashes of light, all repeated one after the other, but with regularity. The formless music of the line, the gliding movement of the unclad gray steel bodies, the routine movements: I can feel myself being gradually enveloped and anesthetized. Time stands still.[7]

The various practices associated with driving cars are similarly "outside time." The postwar period of France's motorization is also the moment when what Henri Lefebvre calls "constrained time" increases dramatically.[8] "Constrained time" is the time of repeated formalities and obligations—obligations that, like the departmental cocktail party, are neither precisely work nor, in any real sense of the term, pleasure. Nothing approximates constrained time better than the space-time of commuting: witness its privileged status for Situationist interventions intent on reinserting historical sensibility into the person most lacking it, the commuter. At the other end of the spectrum of the car's functions from the mundane, daily commute lies the phenomenology of pure speed, celebrated first in France to any mass acceptance by Françoise Sagan, and continuing up to our own day in the panegyrics of Jean Baudrillard. Going fast, as Sagan and Baudrillard both point out, has the effect of propelling the driver off the calendar, out of one's own personal and affective history, and out of time itself. Here is Sagan: "The plane trees at the side of the road seem to lie flat; at night the neon lights of gas stations are lengthened and distorted; your tires no longer screech, but are suddenly muffled and quietly attentive; even your sorrows are swept away: however madly and hopelessly in love you may be, at 120 miles an hour you are less so."[9] And here is Baudrillard, writing in 1967 about cars: "Mobility without effort constitutes a kind of unreal happiness, a suspension of existence, an irresponsibility. Speed's effect, by integrating space and time, is one of leveling the world to two dimensions, to an image; it loses its depth and its becoming; in some ways it brings about a sublime immobility and a contemplative state. At more than a hundred miles an hour, there's a presumption of eternity."[10]

Any initial glance at the intermediate "moment" of the car—its marketing, promotion, the construction of images and markets, the conditioning of public response, the discursive apparatus surrounding the object, in short, its advertising—reveals a discourse built around freezing time in the form of reconciling past and future, the old ways and the new. This is particularly important in a culture like that of France where modernization, unlike in the United States, is experienced for the most part as highly destructive, obliterating a well-developed artisanal culture, a highly developed travel culture, and—at least in the 1950s—a grass-roots national culture clearly observable to French and non-French alike. With such and such a product, the ad reads, traditions, the French way of life, are both conserved and gone beyond; past and future are one, you can change without changing.

But is this really so? Does the paradigm of progress dominate the postwar era in a way that completely erases the disfunctions and suffering inherent to modernization? Even the whisky-soaked voice behind the Sagan myth, the voice that in the 1950s helped commodify a timeless lifestyle based on money, youth, cars and speed, was heard to say quite recently, "I still like speed, but I don't let myself go so much these days. . . . In 1960 you could still drive across Paris, but you'd have to be a masochist to do that now."[11]

Moving Pictures

During the 1950s and 1960s in France mobility was the categorical imperative of the economic order, the mark of a rupture with the past; every individual must be free to be displaced, and displaceable in function of the exigencies of the economic order. The car performed (and continues to perform) the activity most embedded in ideologies of the free market: displacement. It became a key element in the creation of the new and complex image of "l'homme disponible"—Available Man, relatively indifferent to the distances where he'll be sent. Where pre–World War II France treated any demographic movement as a fearful "exodus," and any change in employment as character weakness, post 1950s France-at-the-wheel enacted a revolution in attitudes toward mobility and displacement. It was a revolution that permeated every aspect of everyday life: the automobile at this time became the center at once of a new "sublime" everydayness, a new subjectivity (whose circumference, unlike that of domestic subjectivity, is nowhere and everywhere), and of a new conception of nation. And it was a revolution that saw the dismantling of all earlier spatial arrangements, the virtual end of the historic city, in a physical and social restructuring that matched the transformations of a hundred years earlier. Parisians of the 1960s, like the Parisians of the 1850s and 1860s, saw with their own eyes and lived one Paris intersecting and colliding with another in the process of demolition and reconstruction—a reconstruction that, again like the Haussmanian transformation of the previous century, was completely given over to traffic. During this period the car, as the commodity unlike any other, took center stage in cultural debate; it became the vehicle, so to speak, for dramatizing the lack of real social consensus around the French state-led modernization process, the favorite target of the numerous adversaries of the model of development France had followed since the war. And by a special irony, the expansion of the most vocal and organized of these adversaries, the Poujadist movement of the early 1950s, can in retrospect be seen to be completely dependent on the car.[12] How else could Pierre Poujade, pictured in movement propaganda at the wheel on his countless "tours of France," have been everywhere at once, uniting the *ressentiment* of rural shopkeepers scattered across hundreds of small towns and villages throughout provincial France?

Touring the Renault factory at the dawning of the new age, 1950, author Jules Romains was struck by the vision of "a team where everyone was won over, seized by the same contagion." Trying to evoke the spirit he sensed inside the factory, he lands on a strange analogy: the factory, he says, feels like Morocco under Lyautey. "Suddenly," he writes, "you realize that the contact with such an environment is extremely agreeable. You begin to dream again. You imagine a whole country with the same fever; a country that would self-inject enthusiasm as its daily drug; that would envisage its future not with optimism . . . but with a profound necessity. . . . A whole nation where the joy of working together at a great common project would replace the bitter pleasures of discord."[13] His boosterist tone is more understandable if we consider that Romains was not only celebrating what he saw as the newly nationalized Renault "esprit de corps"; he was also witnessing something like the birth of the modern corporation in France, decades after its maturation in the United States. Thanks largely to the Marshall Plan labor-management "missions" from the United States and the 200 productivity loans of 3 billion francs, the automobile sector and a few others were now in the position to adopt the organizational arrangements, the production and market strategies that would enable them to manufacture for mass markets.[14] The "great common project" that defined France in the 1950s took the shape of what one journalist described as "a vanilla colored toad with four wheels, whose head was exactly the same as its tail."[15] "Une voiture optimiste," whose design was begun under the Occupation, the 4CV was the first French car produced to be affordable on a mass level, the first "people's car" from what had been until then a successful luxury industry, in fact the most successful car industry in Europe. Pierre Lefaucheux, president and general director of the newly nationalized Renault, made the link between this product and the new "modern" France explicit: "I think that the 4CV is indeed in the general line of the evolution of the country and that it must be made as quickly as possible in the greatest possible number."[16] For the first time in the history of French car manufacturing, the marketing of the 4CV was preceded by the creation of a market research service. Production, in fact, could not keep up with demand; in the early years there were a thousand times more buyers than cars. The pride of the 1947 *Salon de l'automobile,* Renault's publicity proclaimed, "It's the car of the year . . . and of all the years to come." Another early ad evoked the alternate twin flagship of French modernization, the bathroom: "The new-born offers all the simplicity of a bathroom."[17] The car was billed as "l'amie de l'homme"—user-friendly, that is, to soothe any anxiety provoked by the intrusion of strange huge machines into one's daily life, and "man's friend" also as a conjugal partner to what were commonly billed as "les amis de la femme": household appliances.[18] The husband, whose preem-

FIGURE 1.2 René-Jacques, Usines Renault, Billancourt, 1951. Courtesy Ministère de la Culture—France.

inence in the family as "the one who works" and thus *must* be motorized, will have one large friend; the wife, on the other hand, will have many small ones: refrigerators, washing machines, electric coffee grinders.

Writing in 1963, Roland Barthes commented that perhaps only food has as much place in French discourse of the period as the car.[19] Barthes's own discursive career had begun in the late 1950s with *Le degré zéro de l'écriture* (1953) and with occasional journalistic pieces (assembled in 1957 as *Mythologies*) on such everyday upheavals as the presentation of the new Citroen Déese on the floors of the *Salon de l'automobile* or the photos accompanying recipes in *Elle* magazine. In 1955 a young sociologist who would soon become the leading theorist of French modernization, Alain Touraine, published his first book, *L'évolution du travail ouvrier aux usines Renault*. Touraine's colleague at Nanterre during the 1960s, Jean Baudrillard, wrote his first book in 1967, *Le système des objets;* it contained a long section devoted to the sublimity of the "fin" of a car and another on the automobile's relation to the world of domesticity. And in 1960, in an article entitled "Situationist Theses on Traffic" published in *L'internationale situationiste,* Guy Debord pronounced the automobile to be at once "the sovereign good of an alienated life and the essential product of the capitalist market."[20]

But when Barthes refers to the saturation of discursive reflection on the car in the late 1950s and early 1960s, he is thinking less of writers like these—less, that is, of writers like himself: the various friends and colleagues of Henri Lefebvre who, following his lead, had turned their attention to everyday life—and more, I think, of the popular information then readily available from magazines, market research surveys, bestselling novels like those of Françoise Sagan, hit movies, advertisements, even documentaries: that array of prose and images that François Nourissier called "the beautiful litany of Mercedes, Facel-Vegas, Aston-Martins and Jaguars."[21] In the early 1950s novelists Boris Vian and Roger Nimier attended the yearly *Salon de l'automobile* as fans; they published their enthusiastic reviews in magazines like *Elle* or *Arts*. Unlike high-culture expositions or events, the *Salon* attracted a decidedly "mixed" audience; it became something of a yearly national festival where rich industrialists from Levallois, farmers from the Ardennes, mechanics from Toulon, Parisian movie stars, and salesmen from Rennes all rubbed shoulders. Françoise Sagan's 1954 car wreck in her Aston-Martin was a focus of print obsession for months, even years; Nimier's latest car accidents were featured in *Femina illustration*. Testimonies to a particular novelist's bold, risk-taking driving style in turn increased sales of both novels and magazines. In the 1950s when a newly resuscitated Céline despaired of having sold less than a tenth of the number of volumes Françoise Sagan had sold, Nimier scolded him ironically for having limited himself to war accidents: automobile accidents brought instant publicity. The violent automobile death of Nimier himself and novelist Sunsiaré de Larcone in 1962, along with those of Albert Camus and Michel Gallimard in 1960, Jean-René Huguenin in 1962, the two sons of Andre Malraux, the Ali Khan, and the near-fatal accident of Johnny Hallyday in the surrounding months, each produced a torrent of horrified, lurid articles. A particular discourse takes place: this is a social fact. But from that point on, the social fact becomes a historical fact. The richness of the reflexive discourse surrounding the car, perhaps more than the car's material ubiquity, points to a particular historical situation.

In fact, the centrality of the car in movies, novels, in the print consciousness of the period to a large extent *precedes* the car's becoming commonplace in French life. As such, the discourse, on the whole, is futuristic: anticipatory and preparatory in nature, fascinated or horrified, but generally permeated with anxiety. In 1961, only one in eight French people (as opposed to one in three Americans) owned a car. And statistics show that the class breakdown of French who bought cars during the period remained constant; despite the arrival of the mass, affordable car, workers, for the most part, went without.[22] Although in the early 1960s

FIGURE 1.3 "This was Roger Nimier's Car." *Parisien libéré,* October 1, 1962. Courtesy Bibliothèque Nationale.

the automobile could no longer be considered a luxury item, it was not yet a banality either. Only in the United States had the automobile been completely integrated into the banality of the everyday to the same degree as the refrigerator or the washing machine—goods whose habitual use effectively removed them from the discursive realm. In France the automobile occupied an intermediate status, that of being *within the purview* of most French. Neither a fantastic, luxurious dream nor a "necessary commodity," an element of survival, the car had become a project: what one was going to buy next. And it is at this point, when the car stands on the verge of becoming a universal accessory, that the cinema no longer represents the car as a fabulous or wondrous item, and the traffic accident loses its inevitability as a plot device in French movies and in the novels of virtually all the popular women novelists of the time.

Postwar French movies chart the demonic nature of the first cars and their metamorphosis into the quotidian. The ability of film—foremost among the media—to create the sensation of the mythic, fabulous object has been heavily discussed in film studies. I want here merely to isolate a number of simple tropes or techniques commonly found in European films of the era that serve to underline the car's singularity; the first of these is the ubiquitous "only car on the road" sequence. Dino Risi's 1962 *Il sorpasso* stars Vittorio Gassman as a compulsive hotrodder who, as he puts it, "only feels good in a car"; for him speed is the mark of an independent and free spirit. The opening sequences show him speeding past a number of familiar Italian monuments in a little convertible sports car. The contemporary viewer is astounded: true, it's the early morning, and later we find out it's a religious holiday—nevertheless, his is the only car on the road in Rome! (Gassman's character, by the way, comes to no good in this movie; the film ends with him killing his repressed law-student sidekick, played by Jean-Louis Trintignant, just after Trintignant has been awakened to the joys of life on the road, by catapulting both the car and him over a sea cliff.) Two extraordinary

French films, Dhery's 1961 *La belle américaine* and Rozier's 1962 *Adieu Philippine,* both deal with ambivalence surrounding private ownership on the part of the first car owners in small, "traditional" communities. Both films choose to emphasize the singularity and newness of ownership by an extended "only car on the road" sequence. The fact that, in each case, half of the village is along for the ride doesn't lessen the sensation of solitude and spatial expansiveness. In one particularly stunning sequence in *La belle américaine* the stationary car becomes unmoored and, driverless, drifts slowly and majestically on its own momentum down a hill, finally coming to rest on a flatboat in the river: an unexpected windfall for the boat's captain. Another common film tactic, more associated with comedy or burlesque realism, is the extreme close-up, such that the car, like some enormous insect on a microscope slide, occupies the whole frame; what little background remains in the frame only heightens the wondrous nature of the object. In the opening sequences of Jacques Demy's 1960 film *Lola,* the camera hugs the luxurious movement of the massive white American convertible—referred to by all the characters who see it as a "voiture de rêve"—as it rolls along the seedy wharfside road in Nantes. Mic's gleaming white brand new Jaguar in Carné's 1958 *Les tricheurs* (the car in which she will die) descends slowly into her view (and that of the audience) on a freight elevator in an auto mechanic's dirty shop. The greatest analyst of postwar French modernization, Jacques Tati—who along with Dhery and Demy specialized in depicting a kind of burlesque malleability in the face of change—is perhaps best remembered for his Busby Berkeley–like pinwheels made of cars doing their intricate synchronized swimming in late movies such as *Playtime* (1967); much of Tati's later work does focus on the car as fatal element in a ballet of seriality and repetition. But in an early film such as *Mon oncle* (1958), the arrival of the fabulous pink and green Chevrolet ("Everything is automatic!") is treated by the camera as a fantastic and singular visitation, as outlandish as the landscape of the house, an autonomous technical object, its garish colors making it

Figure 1.4 *Lola*

FIGURE 1.5 *Playtime*

a factory of fantastic private illusions in addition to a catalog of ready-made tastes, values, and ideas. The moment of singularity in Tati is, however, short lived: turn the corner, and the car has joined an endless row of vehicles in a traffic jam of parents waiting to drop their children off at school. And this was in fact the case: just at the very moment that the car in France is poised to become commonplace, an object of mass consumption, the cinema helps produce a myth or ideology of the car's auratic singularity.

In many of these early films American cars help reinforce the idea of singularity—in fact, the most effective way to indicate an "object from another planet," the effect of intrusion, is to use a foreign, preferably American, car. Claude Lelouch's immensely popular *Un homme et une femme* is in this sense already nostalgic. Made in 1966 when the myths of singularity and speed had been substantially eroded, *A Man and a Woman* shows that they can be nostalgically revived by a Ford Mustang and an actor—Trintignant again—who is no longer an ignorant, technologically wary law student who perishes in a car crash, but a race car driver, a man whose *métier* allows him to place his body in contact with a complex, "souped-up" machine, and whose effective "piloting" of said machine can retrieve the waning provisionary myths of speed and singularity associated only a few years earlier with driving.

The history of driving and the history of moviegoing in France and the United States suggest several curious overlaps and parallels. The foremost manufacturer of automobiles in the world before World War I was France; similarly, France was the world's biggest supplier of films up to 1914, its leading firms, Gaumont and Pathé, dominating the international distribution networks.[23] But the French never recovered from the American jump to superiority in both industries after World War I, a jump in which advances in car production technique stimulated similar advances in film production. Well-capitalized American studios, in other words, were quick to pick up on the assembly line and scientific marketing techniques worked out in the auto industry. As film historian

Figure 1.6 *Un homme et une femme*

FIGURE 1.7 *Un homme et une femme*

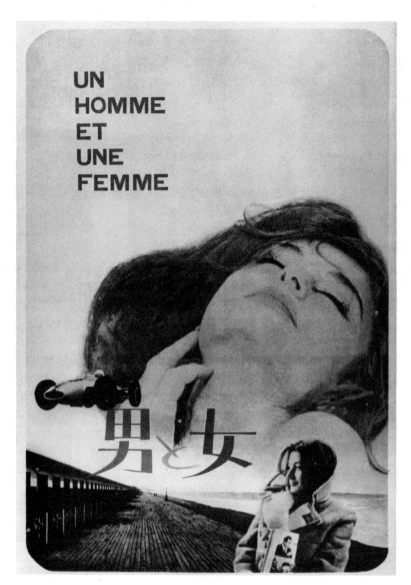

FIGURE 1.8 Japanese advertisement, *Un homme et une femme*

Victoria de Grazia puts it, "During the 1920s and 30s America's movie industry offered an entirely new paradigm for organizing cultural production on industrial lines: what Fordism was to global car manufacturing, the Hollywood studio system was to producing a mass-produced internationally marketed cultural commodity."[24] But her analogy is not quite right, for Fordism is not merely a set of rationalized techniques; its major accomplishment is that of transforming its workers into the consumers of the product they make, and this is certainly not the case with film. And while Fordism introduced techniques, management organization, and the competitive pressure that forced European manufacturers to embark on their own mass production of autos, in the case of film, the American commodity, for the most part, *took the place of* the native product. Another way of putting this is that there were very few American cars actually purchased in France in the late 1950s, but there were an enormous number of American movies watched: by virtue of the Franco-American credit agreement signed in 1946 by Leon Blum and American secretary of state James Byrnes, Hollywood and its European affiliates came to exert a virtually total domination of the European movie market after the war.[25]

The best film historian of this collision of cultures, Dennis Turner, describes the relationship in these terms: "Unlike other American industries which stepped into the vacuum created by the war to sell necessities the indigenous manufacturers could not provide, the American film industry stepped in to peddle a way of life which the native audiences would never quite be able to realize."[26] Although the French were more successful than the rest of the world in stemming the tide of the American cinematic invasion, statistics show that even the Nazis were better for the French film industry than the Americans. By 1947, in other words, American films had overrun the country. The leverage and power wielded by the eight Hollywood studios in the postwar period—a moment when the United States enjoyed unprecedented political, military, and economic superiority over its trading partners—can be measured, as Turner suggests, less in terms of the industry's corporate success than in terms of its role as semiofficial propaganda machine for "the American Way of Life." The postwar screens of Europe were filled with an illustrated catalog of the joys and rewards of American capitalism; all the minutiae of domestic life in the United States, its objects and gadgets and the lifestyle they help produce, were displayed as ordinary—that is, the background or trappings to convincing, realistic narratives. But ordinary or common objects became assertive when they appeared on the European screen. In production, cars had paved the way for film; now, film would help create the conditions for the motorization of Europe: the two technologies reinforced each other. Their shared qualities—movement, image, mechanization, standardization— made movies and cars the key commodity-vehicles of a complete transformation in European consumption patterns and cultural habits.

Much of that transformation involved a change in perception, a change in the way things were *seen*. Writing about the introduction of railway travel and its effects on everyday life in nineteenth-century Europe, Wolfgang Schivelbusch argues that the train, and the accelerated circulation of commodities it both enabled and represented, altered visual perception.[27] As people became accustomed to train travel, traditional perception was replaced by "panoramic perception": the kind of perception that prevails when the viewer sees objects and landscapes through the apparatus that moves him or her through the world. Panoramic perception occurs when the viewer no longer belongs to the same space as the perceived object; as such, it pertains as much to the car driver as to the railway traveler. But in postwar France the banalization of car travel coincided with the banalization of cinema; Georges Perec, for example, locates the historical newness of his young adult characters in the 1960s in terms of their relation to cinema: "Above all they had the cinema. And this was probably the only area where they had learned everything from their own sensibilities. They owed nothing to models. Their age and education made them members of that first generation for which the cinema was not so much an art as simply a given fact."[28] Surely the intensification of two burgeoning technologies, acting in tandem, would produce a qualitative acceleration in panoramic perception; for both cars and movies create perception-in-movement. The automobile and the motion it creates become integrated into the driver's perception: he or she can see only things in motion—as in motion pictures. Evanescent reality, the perception of a detached world fleeting by a relatively passive viewer, becomes the norm, and not the exception it still was in the nineteenth century.

Schivelbusch links the emergence of panoramic perception in nineteenth-century Europe to the physical speed of the train and to the commodity character of the objects it transported. Transportation, in fact, creates commodities: as Marx pointed out in the *Grundrisse,* it is precisely the movement *between* geographical points that makes the object a commodity: "This locational movement—the bringing of the product to the market, which is a necessary condition of its circulation, except when the point of production is itself a market—could more precisely be regarded as the transformation of the product *into a commodity.*"[29] That is, the train coincided with a qualitative change in the production/circulation complex in part by bringing a new level of speed to the circulation of goods. Circulation increases the commodity character of goods with a rapidity that is in direct proportion to the physical rapidity of the vehicles. But what about postwar car traffic? In France at least, the car marked the advent of modernization; it provided both the illustration and the motor of what came to be known as the society of consumption. But passenger cars, unlike trains, rarely transport goods; they transport workers insofar as they themselves constitute commodities. Bicycles were the mode of transportation preferred by workers on the eve of the Popular Front. Postwar modernization and the monopolistic restructuring of industries created the need for a mobile work force, a need identical to the one that characterized the period of industrial conquest in the United States thirty years earlier. The popular car was born of that need. The "moving picture" produced by cars and movies reflects a new acceleration in commodity production and circulation, but it does so, perhaps, far more through a more thorough and complete commodification of the driver, the worker, through a recasting of his identity by means of continuous displacement; in this way man becomes "l'homme disponible." Movable, available man (and woman) is open to the new demands of the market, to the imaginary worlds pictured on the silver screen, and to the lures of the newly commodified leisure of the countryside and the institution of *les vacances,* access to which is provided by the family car.

In the films of Jean-Luc Godard, the new postwar version of panoramic perception then comes to be reproduced, refabricated in such a way that the unconscious relays between moviegoing and driving are put to full use by the director. Referring to the highly artificial sequence in *Pierrot le fou* (1965) showing Anna Karina and Jean-Paul Belmondo full-face in the front seat of a car, "driving" toward the camera as various colored spotlights sweep over them, Godard writes, "When you drive in Paris at night, what do you see? Red lights, green, and yellow ones. I wanted to show those elements, but without necessarily situating them the way they are in reality. More like the way they appear in memory: red stains, green, yellow gleams passing by. I wanted to refabricate a sensation using the elements that compose it."[30] Rather than representing driving, film is used to represent the kind of perception, the blurred sensation, that film and driving have brought about.

Other peculiar conflations of the worlds of driving and movies are woven into industry lifestyles of the era. Youthful director Louis Malle hires novelist and sports-car afficionado Roger Nimier to write the dialogue for his first film, *Ascenseur pour l'échafaud* (1957).[31] "Together [Malle and Nimier] would share a passion for rapid cars and trips improvised in the middle of the night."[32] The completed film featured two adolescents, seduced by automobile speed, who steal a car from a cynical *ex-parachutiste,* home from the colonial wars in Indochina and

FIGURE 1.9 *Ascenseur pour l'échafaud*

North Africa. The car's glove compartment yields the gun the kids will use to commit a murder, as though the accessories and props of a still alien lifestyle (or of American film noir) have the power to condition and determine the gestures of the film's characters. Nimier's only other film endeavor (the screenplay for Jean Valère's 1961 *Les grandes personnes*) turned on a thematics of the car wreck; Nimier himself would die at the wheel of his Aston-Martin in September 1962, several weeks after commenting to Simone Gallimard that "they will find me dead on the autoroute."[33]

Jacques Tati's 1949 comedy *Jour de fête* was one of the earliest popular films to thematize directly the imbrication of movies and traffic speed. (Tati's oeuvre in its entirety supports Schivelbusch's assertion that bourgeois cultural development of the past three centuries is closely connected with the actual development of traffic.) A provincial French postman named François goes to see a promotional modernization film about the postal service in the United States; unstoppable Yankee mailmen are seen diving out of helicopters to deliver the mail. Taking the lesson to heart, François adopts the watchwords of modernization, speed and efficiency, as his own; soon the whole village becomes spectator to François's "doing things the American way!"

What is at stake in any discussion of the Americanization of France is less a conflict between two opposing ideologies than between two different economic organizations within the same philosophical framework. The United States enjoyed tremendous economic advantages in marketing its films in France; as a result, postwar French film directors were forced to develop a perspective from which they could come to terms with the myths of America that had descended so forcefully over French consumers. Thus Tati, the only postwar French director to turn a profit, made films treating the Americanization of everyday life in France that in fact thematized the situation the French film industry found itself in. The use of American icons, stereotypes, and genre conventions by Tati and others was not because of any easy or unam-

FIGURE 1.10 *Jour de fête*

bivalent identification with Americans but because the distance French directors felt between themselves and their own culture was in some ways seen to be the work of Americans. Some older French directors such as Jacques Becker reacted with a form of defensive chauvinism;[34] for younger directors, some of whom felt little or no investment in their national film heritage—including such self-conscious, theoretical film directors as Godard and Truffaut—American cinema was simply the best in the world. As Godard put it,

> When we started making films in France, the French film was much admired. As a reaction against that attitude we used to say, "The commercial American movie is better, in the long run, than such and such a French art film . . . a little gangster film from Hollywood is more important than a French film written by some academic and adapted from André Gide." . . . But that was only a reaction. What was really going on is that we were living under the mythology of the American cinema. . . . And that's not going to change much. Financially the cinema has been American the whole world over.[35]

The direction followed by Godard's own work from the late 1950s to the mid–1960s shows an interesting evolution. All of his films from this period establish a dialogue, directly or indirectly, with American cultural hegemony. But the distance from his first film, *A bout de souffle* ("*Breathless*"), made in 1959, to films like *Weekend* or *Made in U.S.A.*, made six or seven years later, is significant. Around the time of *Breathless* Godard wrote that "things American have a mythical element which creates their own existence";[36] his hero, Michel Poicard, steals only T-Birds and Cadillacs, and worships Humphrey Bogart: the film's style slides unevenly back and forth between Hollywood and Paris as though its director shared his hero's fascination for "things American." *Weekend,*

FIGURE 1.11 *A bout de souffle*

however, best remembered for its eight-minute tracking shot of an interminable car wreck, registers Godard's total estrangement from the American-inspired technological wizardry he had admired in his days as a film critic.

But in the late 1950s young French film directors and moviegoers alike tended to prefer the American product, produced and distributed with assembly-line regularity in tight little genres. For the French, American studio films combined a kind of technological superiority with the representation of a pristine, wild environment; a professional gloss that reinforced narrative credibility with a relatively uncomplicated moral universe. The combination of fantasies of technological sophistication and the wild nobility of wide open spaces is a combination that the car, and the "American way of life" predicated on the car, provided as well.

Mic, the main female character in Marcel Carné's 1958 movie, *Les tricheurs,* remarks, "I wouldn't mind dying like Dean: young, and at great speed." Dead in his Porsche on the desolate road to Paso Robles in 1955, James Dean provided a legend of angst-ridden mobility, a particularly appealing package of the American myth of speed and freedom. The mutinous but self-reliant teenager, at home neither in civilization nor adulthood, who, like Huck Finn before him, "lights out for the territory" had enormous resonance in rapidly modernizing nations such as France or Japan in the years following Dean's death.[37] Carné's character, Mic, gets her wish in the movie. Set awash in the *mal-de-siècle* of the sixteenth *arrondissement*—in one scene disaffected party goers drop lit cigarettes onto the cloth roofs of 2CVs below their balcony—the movie charts the love affair between Mic and Bob. The moral conflict, however, is embodied by two other characters: Mic's older brother, a hard-working, married automobile mechanic ("between the war in Indochina and having to learn car mechanics, I didn't have time to be young"), and Alain, the Nietzschean, libertarian drop-out guru figure to the group of young friends, which, despite considerable financial

FIGURE 1.12 *Pierrot le fou*

FIGURE 1.13 *Pierrot le fou*

FIGURE 1.14 *Weekend*

liberty, charts a restricted and well-trod path from the Café Flore to La Rhumerie in St-Germain-des-Prés.[38] Under the influence of Alain, for whom death is defined as "the trappings of a household, the Sunday drive in the 4CV," Mic sets her sights on a white Jaguar, "une voiture formidable," and engages in a complex set of illegal transactions to obtain it. "When I was your age," comments her brother, "girls were interested in dresses." "Not since the two wars," Mic retorts. After her late-night, possibly intentional, fatal car wreck, the film grants the last word to the doctor who couldn't save her life: "What's wrong with these kids? They refuse life, they won't integrate into the community."

Directed by the venerable creator of the classic "poetic realist" French film, *Les enfants du paradis*, *Les tricheurs* was by far the most popular film in France in the late 1950s. It provoked heated debate across the country; entire magazine issues, including the November 1958 issue of *L'Express,* were devoted to examining the uproar caused by the film, thought by most to be a horrifying yet prescient announcement of the emerging new "youth movement." The most vociferous attack on this view came from communist youth, angered by the idea that the ennui and nihilism of wealthy teens in St-Germain-des Prés could be taken as any kind of general representation of French youth.[39] Still, most movie-goers tended to blur the social distinctions actually present in the film (the working class in the form of Mic and her brother) and responded much more forcefully to the generational division as decisive.

Another popular film of the period, Jacques Dhery's *La belle américaine,* enacts the collision of lifestyles brought on by motorization only to conclude with a particularly happy reconciliation. In this comedy a young working couple in the market for some transportation ends up, completely by a fluke, with an enormous American luxury convertible on their hands. When they bring it back to their *quartier,* the grotesque car fills the entire tiny square; neighbors gather around, children caress it. But the car unleashes a series of disasters: Marcel, the husband, is fired from his factory job both because the car makes him late for work and because his bosses are envious. The couple then attempt to *make use* of the car: Marcel's short stint as a chauffeur is disastrous, and when they next enter the car in an upscale competition for "elegant automobiles," they lose: though they possess the central winning element, they lack the rest of the accoutrements (a well-bred poodle, for instance) with which to ornament the lifestyle the car represents. In one scene husband and wife burst into tears: their "trop belle voiture" has ruined their lives. "I thought I would make all of you happy," says Marcel. "What bothers me most is that we are alone in having a car like this." How can they get their old lives back? Only a car crash can save them: inadvertently, the wife backs the car into the horse-drawn ice-cream cart that provides the livelihood for many of the people in the *quartier,* crushing the cart and disabling the car. The film concludes with a close-up of the car, newly decked out with banners, paraphernalia, and draped with the neighborhood children: the *quartier*'s new "modern" ice-cream cart, *La belle américaine.*

Dhery's film, like Tati's *Mon oncle,* depends on spatial devices to stage its conflict: the juxtaposition of two, nonintegrated universes. Marcel and his wife live in a traditional, "artisanal" *quartier,* almost a village within the city of Paris: what lies beyond their lively neighborhood is foreign turf, how the others live. Everything in the neighborhood is in immediate proximity; at the center is the café, with its elegant old, intermittently functional, coffee machine: the space of work (a bicycle-ride away for Marcel), relaxation, meals, and informal social contact interpenetrate effortlessly in a still-rural style of life characterized by the integration of the outdoors with inside. The *quartier* represents a multigenerational social cell more solid and intricate than the nuclear family, a kind of concentric representation of the world, with the "village" at the center and the café at the center of the village. Into this atmosphere descends what the film emphasizes to be the object that more than any other at the time connotes inequality, and disaster reigns until some method, albeit regional, of taming or appropriating the

Figure 1.15 *Mon oncle*

foreign intruder—integrating it into the lighthearted workings of the village—can be found. The happiest of all outcomes: one can acquire America's good features while avoiding its corruption, one can modernize without losing the national (or regional) identity.[40]

The optimism of Dhery's film flies in the face of the way in which France's motorization was already inscribed in the disappearance of a world: the destruction of such "traditional" societies through the creation of a space/time structured by and for the car. In this regard the distance between French and American experiences of motorization could not be greater. Motorization developed in the United States at almost the same time as the industrial infrastructure and coincided with twentieth-century urbanization. French motorization, interrupted by two world wars and slowed by the economic crisis of the 1930s, was too delayed to be organically inscribed in the development of industrialization and urbanization. Its effect, abrupt and enormous, was most felt in Paris and its surroundings where the volume of automobile traffic climbed drastically after the war. On the eve of World War II there were 500,000 cars in the Paris region; that number doubled in 1960, and doubled again to 2 million in 1965—this in a city whose street surface had increased only 10 percent since the turn of the century.[41] Paul Delouvrier, prefect of the Seine and the Haussman of his day, continued in the line of thinking dominant in Parisian planning since Haussman, according to which the needs of street circulation take precedence over all other urban considerations: "If Paris," he wrote, "wants to espouse her century, it is high time that urbanists espouse the automobile."[42] Delouvrier found a great enthusiast for his ideas in Georges Pompidou, who affirmed, "Paris must adapt itself to the automobile. We must renounce an outmoded aesthetic."[43] The building of the Périphérique, the highway encircling central Paris, billed as the principal and most spectacular renovation realized in Paris in several decades, was begun in 1956; the Right Bank Expressway along the Seine was completed in 1967. By 1972 the Périphérique carried 170,000 vehicles daily and averaged an accident a kilometer a day; it had replaced the old *fortifications* with a kind of permeable wall of traffic that, for the Parisian inhabiting and working within the charmed "inner circle," made the banlieues seem some formless magma, a desert of 10 million inhabitants and gray, undifferentiated constructions, a circular purgatory with Paris—paradise—in the middle. From the perspective within (and the movement inward, inside the French hexagon with the end of empire, inside the newly comfortable domestic interiors, the electric kitchens, the modern bathrooms, inside a new vision of conjugality and an ideology of happiness predicated on the couple, inside the metal enclosures of automobiles) the suburbs were that vague terrain "out there": a décor in perpetual recomposition, a provisory space, or what Perec called "espèces d'espaces."

Cars, Couples, and Careers

By the time that the spatial organization of the city and its surroundings assured the reproduction of a way of life structured by the car, new, albeit degraded, myths had come to replace the old ones of spatial liberty associated with speed, horsepower, and the open road. It is at this time that a typical early-1970s commercial French film such as *Nous ne veillirons pas ensemble* could have its two main characters, played by Marlene Jobert and Jean Yanne, appear during a total of forty minutes sitting in the car, before departure or after arrival. Explanations, crises, admissions of failure, false dreams, reconciliations, the whole gamut of sentimental exchange is condensed into the moments before or after driving. Half of a feature film taken up with images of a car parked, doors open or closed, motor running or not, double-parked or in a forest—each of these differences taking on dramatic significance of an utmost banality. Françoise Sagan, both producer and product of the older myth of speed and sensation, writes now, "These days, solitude is a luxury too. People are rarely alone; they're either at the office or at home with the family.

54

I've heard friends, both married men and women, say: 'Traffic jams? You don't know what you're talking about. They're very peaceful. It's the only time you're alone.' . . . They're alone and free for an hour, bumper to bumper."[44] A clothing salesman in the opening of Chris Marker's 1962 documentary *Le joli mai* agrees: to the filmmaker's question "When are you free?" he responds, "On my way to work"—the only place where he is protected from being yelled at by his wife or boss. The commute—Henri Lefebvre's example of the "constrained time" that strangles everyday life in the contemporary era—has become the respite, the retreat. A miraculous object, the car can compensate for the destruction it has created—it can protect the driver and offer solace from the conditions it has helped create. In this the automobile follows the established order of the capitalism of which it is the twentieth-century emblem; for the established order of capitalism, as Marc Angenot points out, only subsists by repeating "I didn't want that," and by looking around in the disarray it has wrought for the means to restabilize.[45] In the later compensatory myths of the car it is its protected interior space that takes on value, its quasi-domestic (but also anti-domestic) function: a home away from home, a place for solitude or intimacy. As Louis Chevalier remarks, "The pleasure of driving in the city will become, as the city is gradually effaced. the pure and simple pleasure of driving, the automatism of the automobile."[46] With the actual decline in mobility brought on by mass car consumption, the inviolate shell of the car can still provide, though in a weakened form, the liberty from social constraint that speed once promised to provide.

The work of producing the domestic or "intimacy" myth of the car was taken up by women writing in a realist vein for whom the print media—novels, magazines—were of course more accessible than the medium of film. Françoise Sagan's spectacular successs began with the publication of *Bonjour tristesse* in 1954;[47] her career coincided with the rise in the marketing of books in cheap, paperback editions: the first French paperback collection appeared in 1952 from Hachette, twenty years after the appearance of paperbacks in the United States. The emergence of the paperback form occasioned bitter complaint from some circles who proclaimed that Proust should appear only between leather covers.[48] Sagan was not Proust, however. And if she is better known now for a mythology of danger and speed (derived largely from her own near-fatal car accident in April 1954, as much as, if not more than, from the spectacular conclusion to her first best-seller), her writing, along with that of less well known but heavily marketed young female "pulp" novelists, also helped perform the slow integration of driving into the web of everyday, lived emotional relations, particularly those of the couple. In the film version of *Bonjour tristesse,* directed in 1958 by Otto Preminger and starring American actress Jean Seberg, the car (a white American convertible) makes its requisite drop over a sheer cliff into the ocean, but before it does so, it also serves as the site for what little emotional affect the bored teenage narrator experiences: "Already my father was separating from me. I was hurt by his embarrassed face, turning away from me at the table. Tears came to my eyes when I remembered all our old complicities, our gay laughter when we drove home together at dawn through the deserted streets of Paris."[49] Or later:

> At six o'clock we drove off in Anne's car. I loved her car: a huge American convertible that looked more like something out of the advertisements she wrote than like something she would select. It suited me perfectly, with all its shiny gadgets, silent and far from the world. . . . Another advantage was that we could all three sit in front, and never did I feel so friendly toward someone as when I drove with them. All three of us in the front seat, elbow to elbow, giving ourselves over to the same pleasures of speed and wind, perhaps to the same death . . . (pp. 97–98)

In another early Sagan novel, *Aimez-vous Brahms . . .* , a woman's mindless gesture of reaching for the car radio becomes a metonymy for all the repetitions of love in a long relationship: all the small, unthinking gestures, performed in this case in the enclosure of a car, that make up a life lived together:

> In Roger's car, she absent-mindedly switched on the radio. For a second she caught a glimpse of her hand, long and beautifully kept, by the wan light of the dashboard. The veins stood out on the back, campaigning their way to-ward the fingers, mingling in an irregular pattern. Like a picture of my life, she thought, then at once reflected that the picture was a false one. She had a career she liked, a past she could look back on without regrets, good friends. And a lasting relationship. She turned to Roger:
> "How many times have I done this before—turned on your car radio as you've driven me off to dinner?"[50]

Months later, long after Roger and Paule have separated, Roger repeats her banal gesture and in so doing conjures up the enormity of the emotional loss he has suffered: "For a few minutes he drove in silence, frowning to himself; then he reached for the radio and remembered. Cherish, he thought, cherish—that was Paule and me. Life had no flavor for him. He had lost her" (p. 101).

In two novels from the mid-1960s, Christiane Rochefort's *Les stances à Sophie* and Simone de Beauvoir's *Les belles images,* the automobile also serves as the emblem of a marriage: its myth and its dissolution. The central event in both novels is a car wreck that fractures husband and wife into two irreconcilable types: the technocrat husband, on the one hand, and the denouncer of ideological manipulation/humanist wife on the other. Both novels are explicitly concerned with capturing or repro-ducing the language of contemporary, new "technocratic" France as spoken by its leading representatives: the prosperous new professional middle classes, or *cadres*. Beauvoir described her project as such: "I turned back to an earlier project dealing with this technological society. It is a society that I keep as much as possible at arm's length, but nevertheless it is one in which I live—through papers, magazines, advertisements and radio it hems me in on every hand. I did not intend to take certain given members of this society and describe their particular experiences; what I wanted to do was to reproduce the sound of it [*son discours*]."[51] Beauvoir and Rochefort were not alone in their focus on a palpably new, historically significant discursive reality and their almost ethnographic relationship toward it. Their project recalls Henri Lefebvre's anecdote about his "discovery," around the same time, of the concept of everyday life. The concept occurred to him, he says, when his wife walked into the apartment holding a box of laundry soap and said, quite seriously, "This is an excellent product."[52] And the quasi-documentary turn taken by Rochefort and Beauvoir in these novels bears a distinct resemblance to that other semi-satire of consumer coupledom, Georges Perec's 1965 *Les choses,* a novel that sets out to explore, as Perec himself put it, the way "the language of advertising is reflected in us."[53] But where the women novelists and Lefebvre highlight the unequal way that everyday life sits on the shoulders of men and women, Perec allows no gender distinction to separate his husband and wife, united by shopping into an intractable third-person plural pronoun: "They." And Perec makes the counterintuitive decision to represent contemporary discourse nondiscursively: not a single line of dialogue is uttered in the novel. Rochefort and Beauvoir, on the other hand, construct their novels almost entirely out of dialogue: dramatized exchanges, general chatter, drifting conversations overheard with half an ear, conversational clichés that spring unbidden to characters' lips. Before beginning to write her novel, Beauvoir compiled a Zolaesque notebook, "as dismaying as it was amusing,"[54] of the discourse of technocracy around her: a kind of *Dictionnaire des idées reçues* of the clichés of modernization, garnered from the media, and from "authorities" like Louis Armand, the great celebrator of Americanization who was responsible for modernizing the French railway system after the war, and Michel Foucault. Rochefort was also drawn to the dictionary form, but she assigns the task of its compilation to her narrator, Céline, a working-class young woman who somewhat inadvertently finds herself married to a technocrat, Phillipe, and who charts the discursive impasse that is her marriage in a "Célinian-Phillipian" dictionary. Céline's dictionary project takes shape slowly with her growing awareness that the people of her postmarital milieu (her mother-in-law, for instance) speak in the precise syntax and tone of advertisements: "'Hand-embroidered cloth is absolutely charming,' says Mme. Aignan. 'Especially for a bedroom. But if you want a more unified look, you have the nylon veil, which is much easier to wash and hardly ever needs ironing.'"[55] In another entry from her dictionary Céline records the effort made by Phillipe's friends to personalize what is commercialized speech by the addition of the sign of subjectivity, the omnipresent—in French at least—"Moi, je . . .": "Example: 'I myself [Moije] find that the Victory is the best car now on the market,' said Jean-Pierre Bigeon, who by the way doesn't know how to say anything else. 'Its compression . . .'—for what follows in exact detail consult the car-manufacturer's brochure. What's remarkable is that once they've got the 'Moi je' out of their mouths, they repeat word for word what they've stolen from elsewhere: content, syntax and vocabulary" (p. 120). Although men and women are equally given over to "ad-speak," the examples I've cited show the rigid gender division of territory that defines conjugality for Rochefort: the domestic interior of double-lined curtains and nylon veil for women; technology and beyond for men. Together the two add up to the newly vitalized unit of consumption energy that is also center stage in *Les choses:* the couple, or *jeune ménage* (an entry in Céline's dictionary).

For Rochefort the impasse between the worlds within the couple is insurmountable; in her fiction generally, if the car is "l'amie de l'homme," then it must be "l'ennemi de la femme." One of her earlier novels, *Les petits enfants du siècle,* deals with a working-class couple living in an HLM (government-subsidized, low-income housing) on the out-skirts of Paris. Having succumbed to the material incentives of the government's postwar "family allocations" plan to increase population through subsidizing childbirth, the couple is shown trying to improve its standard of living by having baby after baby ("'And my fridge is right here!' proclaimed Paulette, tapping her stomach in front of the other women").[56] In this world the car and its discourse are men's domain: "Thanks to [the birth of] Nicholas we could get the washing machine repaired . . . we could get back the TV. . . . After that, with a little luck, we could maybe think about a car. That was what they had their eye on now, more than the fridge, Mom would have wanted a fridge, but Pop said that it was his turn this time to have some comfort, not always his wife's" (pp. 20–21). If the pleasures of speed and solitude are foreclosed from the outset for the 4CV-driving, working-class father of numerous children, they can be replaced by an equally hard-driven discursive competition. In the vacation compound where the family is joined by similar families, women and children discuss children, while the men indulge in a violent display of one-upmanship on the topic of cars. Rochefort devotes six pages of savage parody to the father's con-versation with "the other guys who were also fountains of science; you couldn't shut them up on any subject, they talked about everything with authority, each one trying to show the other he wasn't an asshole, and that he knew something, especially about cars—they always came back to that in the end" (p. 63). If the father in the novel is handicapped by his practical car and his huge family, he nevertheless prevails in the conversational drag race: "he had won the conversation race and that was what counted" (p. 66).

The prosperous male characters of Rochefort's 1963 novel, *Les stances à Sophie*—Céline's husband Philippe and his friend Jean-Pierre—don't need to have recourse to a conversational replacement for racing;

as *planificateurs*—Philippe works at "Decentralization" and Jean-Pierre at "Regroupings"—together their function is to move industries out of Paris to underdeveloped areas in France. They can afford fast cars and more upscale vacations. En route to the seaside, the speeding Jean-Pierre, executing a dangerous pass, crashes into a 2CV, killing his wife (with whom Céline is having an affair) and injuring a child in the other car. From this point on, Céline's disillusionment is complete: she sleeps with an Italian peasant (the rural Italian south, the novel implies, still produces real men; "virility seems to recede as urbanization rises" [p. 106] is how Céline puts it; or, "as soon as you get out of the world of the bourgeoisie, you begin to notice men again" [p. 236]), and the novel concludes with her (literally) walking out of the marriage.

Rochefort is one of the most unequivocal voices of the period at work debunking the way in which the state-led modernization drive is ideologized. Her novels register its costs, costs that the working class and women—or at least wives—of all classes are made to pay. Read today, her work can almost be taken as a textbook illustration of sociologist Luc Boltanski's theory that, in France at least, the race to appropriate road space is largely reducible to class conflict not perceived as such. Driving, in other words, is a more or less direct reflection of social organization; ideology can be traced in the various imposed comportments such as driving, the links created between individuals and society by the merchandise created by that society. The vast majority of road accidents involving men, according to Boltanski, can be analyzed as the result of races between drivers who try "to maximize their gains in space, which would be equivalent to maximizing their profits in time."[57] By choosing as her narrator a class outsider and witness to the discourse of technocracy, and an only temporary resident in the "jeune ménage," Rochefort makes men alone representative of the machinations and ideologies of their class.

Simone de Beauvoir's more complex novel, *Les belles images,* undercuts any simple gender opposition between the two types that in

Rochefort's work make each other real: those who implicitly valorize technology in all cases as a response to natural needs (men), and those others, women, who see in technology only a pretext for manipulation, with the desired ideological effect of "normalization." Beauvoir was at pains to distinguish her position from the one she attributed to Rochefort, one of more or less Luddite technophobia: "Notice that I am not at all in revolt against technical progress in and of itself the way Christiane Rochefort is. I like jets, beautiful machines, hi-fi sets."[58] Laurence, the young wife in *Les belles images* from whose perspective the story is told, is a fairly acute analyst of ideological manipulation, but well she should be—she herself is a successful advertising executive and, as such, complicit in the very production of all the beautiful images that surround her: "I am not selling wood paneling: I am selling security, success, and a touch of poetry into the bargain."[59] Beauvoir explained why she chose such a narrator: "In order to display it [the discourse of technocracy] I had to stand back and view it from a certain distance. I decided to look at it through the eyes of a young married woman sufficiently in agreement with those around her not to sit in judgement but also sufficiently honest to feel uneasy about her complicity."[60] Though the novel, her first in twelve years and certainly her best, sold over 120,000 copies and was on the best-seller list for twelve weeks, it was also the target of critics who denounced its subject matter as inappropriate for Beauvoir; they accused her of making an incursion into Sagan territory, and an inferior one at that. Certain similarities do exist. Like Sagan's adolescent narrators, Laurence can experience intimacy only when in a car:

> A couple. Laurence looked attentively at Jean-Charles. She likes driving, sitting next to him. He is watching the road carefully, and she sees his profile, that profile that stirred her so, ten years ago, and that moves her still. Seen from the front Jean-Charles is no longer quite the same—she no longer sees him in the same way. He has an intelligent and lively

face but it looked—what was the right word?—set, like all
other faces. Seen from the side in the faint dusk, the mouth
seems less decided, the eyes more dreamy. This was how she
saw him eleven years earlier, this was how she saw him when
he was away, and sometimes, for the odd moment, when
she was driving in the car next to him. They fell silent. The
absence of words was like a secret agreement. (p. 24)[61]

But despite certain thematic overlaps with Sagan, Beauvoir's concerns
in *Les belles images* are actually closer to those of Lefebvre. The questions
that preoccupy Lefebvre and Beauvoir are not posed by Sagan's work.
Does the postwar modernizing process act to suppress the memory of
the suffering and destruction that must inevitably accompany it? How
can the individual trapped within the "practico-inert"—what Lefebvre
called "everyday life"—gain some situational autonomy? Beauvoir con-
structs the environment of Laurence's milieu with uncharacteristic hu-
mor out of a relentless use of stock expressions, clichés, and stereotypes,
some of which her narrator can't help but use herself: "She smiles. He
doesn't have many faults, all in all; and when they are driving, side by
side, she always has the illusion—though she wasn't given to traps of
this sort—that they are 'made for each other'" (p. 108). The conflict in
the book is embodied by two characters, Laurence's father and her
husband, Jean-Charles, with an Oedipally transfixed Laurence standing
squarely between the two. The conflict between the two men is essentially one of temporality and lifestyle; it emerges in their differing responses to a shared problem: How can one best avoid the knowledge
of the social costs of one's own privilege in the present moment? By a
retreat into the past or by flinging oneself into the future? Laurence's
father, something of a bohemian—he has no hi-fi system but does own
a collection of beloved classical records—believes that "man has lost his
roots"; he admires the "austere happiness" (p. 200) of dirt-poor Greek
peasants. Jean-Charles, on the other hand, becomes lyrical when he

evokes the techno-perfected future: deserts covered with wheat and tomatoes, all the world's children smiling. Laurence tends to side with her father but live like Jean-Charles: "Jean-Charles is already living in 1985, and Papa is looking back sadly to 1925. At least he talks about a world that did exist and that he loved: Jean-Charles invents a future that may never come into being at all" (p. 50). But a psychological crisis suffered by her daughter brings Laurence face to face with the present and with the distance that separates her from her husband. This distance becomes all the greater after a car wreck in which Laurence swerves to avoid a bicyclist and totals the car; Jean-Charles accuses her of costing him 800,000 francs to avoid the risk of injuring the boy. Laurence cannot, however, throw herself wholeheartedly into her father's world of nostalgia: a longed-for car trip to Greece with her father reveals peasant life to be austere, yes, but poor above all and not particularly happy. Alienated from both her father and her husband, Laurence at the end of the novel stands on the brink of carving out not freedom exactly, but a more autonomous stance.

Beauvoir wrote that by tracing Laurence's simultaneous disillusionment with her father and her husband, she wanted to show that the two kinds of bourgeoisies—the old, traditional one of the father, which believes in absolute moral values, and the new technocratic one, with its revised values of efficiency, competence, and professionalism—were actually one and the same. Both the older style bourgeois and the new "relaxed" style—the kind that read L'Express and had no time for elaborate gastronomy—were engaged in massive myth production and collusion to deaden the realization of the existence of a huge population left out of the modernization process. But the novel was far more successful at parodying the new "techno"-bourgeoisie, and this led many readers to identify Beauvoir's own position with that of Laurence's father: nostalgic and passéiste. Beauvoir herself was not, in fact, an Ellulian technophobe like the character she created; her biographer, Deirdre Bair, reports that to console herself after the break-up of her long affair with Nelson Algren, Beauvoir decided to, as she herself phrased it, give her "love" to another—her new car: "Gallimard loaned her the money to buy it, Genet helped her find it, and she was now the proud owner of a new Simca Aronde. She was a terrible driver and in years to come would be involved in many minor scrapes and several serious accidents."[62] Her Simca notwithstanding, Jean-Jacques Servan-Schreiber, co-founder of *L'Express* magazine, chose Beauvoir to be his example of the passé, elitist "intellectual of the older generation" in his book *Le défi américain,* perhaps because Beauvoir had explicitly attacked *L'Express* in the pages of her novel. Through her main character, Beauvoir gets in a hearty jab at the management and ideology of the magazine: "Laurence opened *L'Express:* the news, dealt out in slim lines, could be swallowed like a glass of milk—no roughness, nothing that stuck, nothing that rasped" (p. 124). For Servan-Schreiber, Beauvoir is a perfect example of

> [those] who speak of the excesses of consumer society [but who] are really attacking the consumer's right to determine his own needs. The condemnation of this small, but precious, aspect of economic democracy, indicates a resurgence of "enlightened despotism." An elite convinced of its own wisdom is certain it has the right to impose its own preferences through specific restraints. It would even be ready, so it says, to restore poverty in order to protect the masses against the moral risks of growth in an atmosphere of freedom.
> Let us listen to Simone de Beauvoir . . . [63]

Servan-Schreiber then goes on to quote a particularly romanticized speech made by the character of Laurence's father on the superiority of nontechnologized communities in Sardinia.

Written a year after *Les belles images, Le défi américain (The American Challenge)* quotes heavily from the same experts, such as Louis Armand, whose prophecies and pronouncements Beauvoir collected in her note-book and placed into the mouths of her technocrat characters. *Le défi,* which cast itself "not as a history book but, with some luck, a call to action" (p. 278), criticized France for continuing to suffer progress rather than pursue it. Servan-Schreiber advocated France's salvation through a discriminating Americanization, and in particular through its embrace of an American, flexible managerial style: "Americans are not more intelligent than other people. Yet human factors—the ability to adapt easily, flexibility of organizations, the creative power of teamwork—are the key to their success" (pp. 251–252).

To debate the accuracy of Servan-Schreiber's casting of Simone de Beauvoir in the role of despotic older statesman/intellectual would be less interesting than to consider his stakes in doing so. At issue is a kind of changing of the guard: the replacement of the reigning postwar intellectual couple, Beauvoir and Sartre, by the new reigning couple of "everyday life": Servan-Schreiber and his coeditor at *L'Express,* the woman whom, as she herself puts it in her memoirs, Servan-Schreiber "to a certain extent, invented"[64]—Françoise Giroud. Sartre's domination of the postwar period is well known; in fact, the survey of young people taken by *L'Express* in 1957 showed Sartre to be the intellectual thought by a large majority of those surveyed to have most influenced his era. In 1954 the publication of Beauvoir's *Les mandarins* (crowned by the Prix Goncourt) marked the peak of the couple's legendary status. But the battle between the two couples and what they represented spanned some fifteen years from the early 1950s to the mid-1960s; it can be described in a number of ways: as the war between a journal, *Les temps modernes,* and the first French mass-circulation news weekly, *L'Express;* between a humanist leftism for which "the people" and "praxis" were relevant, indeed central, terms, on one side, and a politics that called itself Left but that had accepted market economy in its entirety and as such had stepped outside of any conventional division between Left and Right, on the other; between the Sartrian camp's rhetoric of anti-"American imperialism" and the *Express* crew's belief in the use of American capital, technologies, and symbolism to transform, if not the substance, at least the style in which political power appeared.[65] If Giroud and Servan-Schreiber never attained the status of "great intellectual couple" that Sartre and Beauvoir possessed in the mid-1950s, it is in part because their reign announced the waning of such a category.

That Giroud and Servan-Schreiber nevertheless emerged the victors of the long war can be gauged by their success in having the battle be fought at the level of style—whether this be the austere prose style of *Les temps modernes* versus the seductive, newly Americanized journalistic format of *L'Express;* or the image of Beauvoir, described by Giroud in an early portrait as "massacreing a handsome, intelligent face with the hairstyle and accessories of a fishwife decked out for Sunday mass;"[66] or the image of Sartre and Beauvoir in her Simca compared with that of Servan-Schreiber (described by Arthur Schlesinger in his preface to the American translation of *Le défi américain* as "a European of the Kennedy generation and style" [p. xii][67]) with Giroud at the wheel of her little white American convertible. Giroud's affection for her convertible was in fact only matched by the animus of her rivalry with Simone de Beauvoir; in Giroud's memoirs she comments that given the number of significant intellectual and political seductions she accomplished while driving, her car should have been classified as a historical monument. François Mauriac, until then notoriously prickly and standoffish, melted

one August 15th when I drove him from Megève to Paris. He was supposed to get a ride with someone else. And then well, there he was, one bright morning, sitting calmly in my car, a little American convertible. . . . And we were off. When you drive together, the route creates an intimacy. In a few hours of travelling and a lunch, you can sometimes end up saying more things than you might in several years. . . .

In the same car I spent a summer day with François Mitterand. He said to me 'Let's go for a ride' and we took off without a plan in our heads.[68]

In the years immediately following the inauguration of *L'Express* in 1953, enough shared ideological views enabled a group of intellectuals as diverse as Camus, Mauriac, Merleau-Ponty, Sartre, and even Malraux to write for the magazine. Sartre chose *L'Express* to publish the article announcing his definitive rupture with the *Parti communiste français*. But even then, according to Giroud, the war was underway: "Even though my personal relations with Sartre had always been very pleasant, it was certain that I wasn't loved in that camp. It must be said that at that time Simone de Beauvoir accused Maurice Merleau-Ponty of being 'a bourgeois betraying the cause of the people' because he too had decided to write for *L'Express*."[69] Giroud's most extended criticism of Beauvoir echoes that of Servan-Schreiber in *Le défi américain;* detached from the lot of "everyday women," Beauvoir nevertheless seemed to prescribe a set of behaviors for women to follow:

> In the past I was irritated by the positions that Simone de Beauvoir took, not because of what she said, but because the woman who was urging others to assume their freedom was herself assured—on the material plane—of a professor's salary and—morally—of the unfailing support of a man. "I knew that if Sartre agreed to meet me in two years on the Acropolis, he would be there. . . . I knew that he could never do anything to hurt me . . ." Who wouldn't opt for that kind of freedom? But it's what I would term "snug freedom": material security. It's artisanal work, protected and guaranteed, if I may say so! . . . Simone de Beauvoir's personal life is beautiful, because there is nothing rarer than a successful human relationship, but nonetheless, as it relates to women

in general, it's a kind of involuntary imposture, since she cannot say "Do as I do." She is the example incarnate of a woman living for and through a man, a woman who, if I read her rightly, never really had to give up anything in her relationship with that man because it intruded on her work.

It is probably an enviable success, and one devoutly to be wished. But strictly speaking it is inimitable, a success which does not add constructively to the life of "Everywoman." And the lives of everyday women are too difficult for anyone to have the right to lie to them. You do not become Simone de Beauvoir-hyphen-Sartre because you make up your mind to have an abortion and to live together without being married, any more than you become Bridget Bardot because you use the same beauty cream she does.[70]

For Giroud the working and affective conditions of Simone de Beauvoir within her privileged couple amount to "protected" or "guaranteed" labor, that of a pampered craftswoman in a guaranteed market, an entitlement no longer relevant in the era of the new, competitive, mass or "everyday" woman. Sartre and Beauvoir spoke in the name of the people from Olympian heights, prescribing how they should live; Servan-Schreiber and Giroud spoke the language of the new bourgeoisie-without-blinkers, anticipating how they desired to live. The offices of *L'Express* itself had already been transformed from its family-style, artisanal workshop into a factory of uniform production.[71] Gone was the approximate, unpredictable, but above all disorderly format of the magazine; had Sartre wanted to write for *L'Express* after 1964, he could not have. Its new format, borrowed from *Time,* featured short, unsigned news chunks in a standardized, accessible style; articles, no matter what the subject matter, were to be made in the same way, following assembly-line principles and containing the same ingredients: the typical detail, the brief citation, the amusing anecdote. The procedure was simple: a reporter gathered information and sent back notes; someone with the necessary meticulousness extracted the essential; someone else did the composition, and a document checker verified figures and names. Servan-Schreiber hung a huge sign in his office proclaiming (in a striking echo of contemporaneous developments within structuralism) that "'I' doesn't exist."[72]

Servan-Schreiber's *Le défi américain* sold more copies in the three months after its publication than any book, fiction or nonfiction, published in France since the war. The circulation of *L'Express,* which had fallen after the end of the Algerian War, climbed to new heights. By 1966 the ascension of Servan-Schreiber and Giroud to the status of reigning everyday-life couple was one mark of the end to the period of debate surrounding the state-sponsored march into modernization; France's cultural and psychic assimilation of the car and the new geography it had created was another. For the moment, at least, a shared representation of French interests and identity prevailed.

Hygiene and Modernization

Man's Voice: "The new Alfa Romeo . . . with its 4-wheel disk brakes, luxurious interior, and road-holding ability, is a first-rate "*gran turismo*": safe, fast and pleasant to drive with quick get-away and perfect balance."

Woman's Voice: "It's easy to feel fresh. Soap washes, cologne refreshes, perfume perfumes. To combat under-arm perspiration I use Odorono after my bath for all-day protection. Odorono comes in spray-bottle aerosol (it's so fresh!), stick or roll-on.

—*Godard,* Pierrot le fou *(1965)*

Housekeeping

In attempting to account for the frenetic turn to large-scale consumption in postwar French society, a popular biological metaphor prevails: the hungry, deprived France of the Occupation could now be sated; France was hungry and now it could eat its fill; the starving organism, lacking all nourishment, could gorge on newfound abundance and prosperity. In this quasi-ubiquitous narrative of wartime deprivation, France appears as a natural organism, a ravenous animal. That its inhabitants should in a very brief time completely alter their way of life and embrace a set of alien habits and comportments determined by the acquisition of new, modern objects of consumption is seen to be a *natural, necessary* development. The following almost randomly chosen passages from French memoirs that span the period show both the necessity of the cliché—there is no way to talk about postwar France without relying on it—and its gradual evolution: from a literal hunger for food to a more general appetite for consumption per se. Alphonse Boudard's account of the postwar atmosphere of 1946 sets the scene: "And then always, now, for six years, these eternal questions of food [*bouffe*] . . . the ration cards . . . the meat, the milk, the cooking fat missing from the frying pans of a France, liberated, but with an empty stomach."[1] Reminiscing about the immediate postwar days in Paris, Françoise Giroud writes,

> And anyone who wasn't in France in those days cannot understand what it means to be hungry for consumer goods, from nylon stockings to refrigerators, from records to automobiles—to buy a car back then you had to get a purchase permit and then wait a year. . . .
>
> It's very simple: in 1946 in France there was literally nothing.[2]

In a recent autobiographical work, François Maspero speaks of the effects of wartime deprivation: "For a long time . . . the child of the war that he once was had lodged inside of him a tiny tenacious fear: the haunting worry, anchored in a corner of his memory, that once again everything might *come to a stop*. Because he had known days when there wasn't any gas, any central heating, no more electricity or hot water. Days when there was nothing to eat. When *things* were absent, soap or socks."[3] But such explanations are themselves already part of an ideology of consumption that is now invoked to conceal the more complex, *unnatural* causes of the abrupt postwar French turn to American-style mass-consumption habits. In 1956, in a short piece on skin cream he would include in his *Mythologies* published a year later, Roland Barthes makes use of an ideologeme already prevalent in French discursive reality—one whose elaboration, I believe, will take us much farther than the "hungry France/sated France" narrative. If France is hungry, Barthes suggests, it is neither for food nor for the things whose existence French children of the war, such as Maspero, now found so precarious; its deep national psychic need, which he names but does not analyze, is to be *clean:* "'Decay is being expelled (from the teeth, the skin, the blood, the breath)': France is having a great yen [*fringale*] for cleanliness."[4]

Fringale in French can mean either a pressing, violent hunger or an irresistible desire: France is hungry for purity, it yearns for, demands to be clean. *Mythologies,* with its essays on laundry detergent and semiotic analyses of bleach, its hermeneutics of skin hygiene (depths and surfaces), and its dazzling conjuring up of the smooth, streamlined gloss of the latest model Citroen, is not alone in its isolation of a qualitatively new, French, lived relationship to cleanliness. A glance through the other books of the time engaged in examining lived, social reality (those other early chronicles, with Barthes, of the everyday) reveals a striking fact: when each of the authors turns to a discussion of the new role played by advertising in postwar society, he uses as his primary example advertisements for laundry soap. Lefebvre, as I mentioned earlier, attributes his whole discovery of the concept of "everyday life" to his wife's tone of voice, one day in their apartment, when she praised a particular brand of laundry soap. Baudrillard in *Le système des objets* devotes a lengthy section to an analysis of a Pax detergent commercial, an analysis that enables him to develop a general theory of advertising that he would go on to expand in his 1966 *La société de consommation.*[5] Yet despite this symptomatic return to the *example* of soap, to soap as an example, none of these writers goes farther than noting, as Barthes does in his broad yet elliptical generalization, that France is undergoing a massive desire to be clean. No one, that is, offers any explanation for it. What is the relation between cleanliness and modernization in postwar France? Why would such a national desire express itself at this historical moment? How does the culture of cleanliness contribute to a new conception of nation?

"France was being regenerated, it was being washed of all the stains left behind by four years of Occupation."[6] Certainly the immediate postwar purges (called *épurations* or "purifications") and attempts to rid the nation of the traces of German Occupation and Pétainiste compromise and complicity set the tone for a new emphasis on French national purity. Historian Robert Paxton is unable to avoid a vocabulary of moral stain when he describes the postwar purges, the process whereby collaborators and those who had compromised with the Vichy regime were punished and removed from positions of authority: "Officially, the Vichy regime and all its works were simply expunged from history when France was liberated. . . . [*But*] for good or evil, the Vichy regime had made indelible marks on French life."[7] Some stains, in other words, you can't get out. Also in the years immediately following the war, while "the *tribunaux épurateurs* were working day and night," Marthe Richard, a municipal councillor in Paris (and one of the first women pilots), was launching another social hygiene campaign, that of closing the brothels in France: "Moral cleanliness! Purification. . . . Pull out the evil by the root!" Once the 177 brothels in Paris were closed down, Mme. Richard called for the next step: "the mopping-up [*nettoyage*] of the streets and the sidewalks"; in December of 1945, in a Declaration to the Conseil municipale, Mme. Richard declared, "The moment has come to propel ourselves toward the goal of cleanliness and moral progress."[8]

It does not seem to be moral progress that Alain Robbe-Grillet had in mind in the mid-1950s when he wrote the short essays that, published together under the title of *Pour un nouveau roman,* served as a kind of manifesto for change in contemporary French high literary production. But he too, like Mme. Richard, appears to be engaged in a project of redemptive hygiene. Read today, what is most striking about Robbe-Grillet's propositions for the novel is the energy with which he proposes to clean the Augean stables of the realistic novel form of the fetters and archaisms that "keep [us], ultimately, from constructing the world and the man of tomorrow."[9] The goal, for Robbe-Grillet, is to arrive at a prose form capable of representing the new, depthless here and now, the era of the masses, which is "one of administrative numbers," and no longer, like the earlier period of high realism associated with the figure of Balzac, marked by the rounded, individual character, no matter how typical (p. 29). The novelist, for Robbe-Grillet, must be eternally vigilant, on the lookout for the tell-tale stains of an outmoded romanticism that lurk in the form of animistic descriptive adjectives and metaphors: "Man and things would be cleansed of their systematic romanticism" (p. 39). All projections of depth—which is to say, of human significance—must be eliminated in order to arrive at the picture of a world that is "neither significant nor absurd. It *is,* quite simply. Around us, . . . things are *there.* Their surfaces are distinct and smooth, intact" (p. 19).

The cleansing process must be thorough: "Nothing must be neglected in this mopping-up operation [*entreprise de nettoyage*]" (p. 57). But how is it to be accomplished? First, by a thoroughgoing and determined cleansing of literary language: the novelist must strip away visceral adjectives, metaphors, any analogical or empathic trope that renders the world of objects tragic or conductive of any human significance whatsoever. Meaning—for Robbe-Grillet an extraneous and anthropomorphic addition—is a useless excess: "An explanation . . . can only be *in excess,* confronted with the presence of things" (p. 40). When this fundamental cleansing of figurative language has been accomplished, then "the world around us turns back into a smooth surface, without signification, without soul, without values, on which we no longer have any purchase" (p. 71): a world of desire without values.

But how does one go about redeeming literary language from the polluting propensities of metaphor? The answer, for Robbe-Grillet, is simple: by trusting in "the cleansing power of the look" [*le pouvoir laveur du regard*] (p. 73). The novelist must rely entirely on the sense of vision ("in spite of everything, our best weapon"), but this new kind of visionary has none of the fanciful, impractical, or speculative qualities traditionally associated with the term. Robbe-Grillet himself trained to be an agricultural engineer, and the new seer shows the traces of such a formation: his visionary activities have been stripped down, his vision itself cleansed and focused to become a tool for conducting a set of technical, almost administrative operations based on criteria of efficacy. The "cleansing power of the look" is a humble power that limits itself to merely measuring, locating, limiting, defining, and inspecting. But this humble power is in fact a far-reaching one. In a reading of *La jalousie* Jacques Leenhardt situates such "morbid geometrism" in the ambiance of a waning colonialism, arguing that the activity of inspection, of obsessive visual surveillance, that dominates the novel constitutes the repressed situational context of the colonial situation: "The right to look without being looked at," he argues, is a microcosm of the colonial problem.[10]

Yet within this generalized postwar atmosphere of moral purification, national cleansing, and literary laundering, journalist Françoise Giroud could still cause, in 1951, what she later called the only scandal she involuntarily provoked, by publishing an investigation/survey in the recently launched women's magazine *Elle,* entitled "La Française, est-elle propre?" ["Is the French Woman Clean?"]. Perhaps certain people (Germans) had left a polluting stain on France, perhaps certain French (collaborators) had to be purged and eliminated, perhaps certain French women (brothel owners and prostitutes) were tainted, perhaps literary language was hopelessly metaphorical and in need of a good scrubbing, but to question the personal hygiene of *la Française*—the French woman? "I'll admit that investigation was really meant to provoke," writes Giroud, denying any moralizing motivation: "When dealing with something like cleanliness, it was interesting to tell women the truth. 'You buy a dress because you want to look good, to please, but under your dress what are you wearing? A garter-belt (pantyhose didn't exist) that hasn't been washed in two years. That's the national average. So don't go around scolding your child because he doesn't wash his hands before sitting down to meals. You're the one who's dirty.'"[11]

The historical record can be expunged, the foreign occupier driven out, the morally diseased or tainted elements of the national body cleansed or surgically removed, but to target a nation's women? This— as Frantz Fanon said around the same time a propos of France's own campaign to colonize Algeria according to the well-known formula "Let's win over the women and the rest will follow"—is to target the innermost structure of the society itself.[12]

To evoke the colonial situation here is not gratuitous; I want to suggest that in the roughly ten-year period of the mid-1950s to the mid-1960s in France—the decade that saw both the end of the empire and the surge in French consumption and modernization—the colonies are in some sense "replaced," and the effort that once went into maintaining and disciplining a colonial people and situation becomes instead concentrated on a particular "level" of metropolitan existence: everyday life. (This is what is meant by the capsule phrase "the colonization of everyday life," proposed by the Situationists and by Henri Lefebvre at the time.) And women, of course, as the primary victims and arbiters of social reproduction, as the subjects of everydayness and as those most subjected to it, as the class of people most responsible for consumption, and those responsible for the complex movement whereby the social existence of human beings is produced and reproduced, *are* the everyday: its managers, its embodiment. The transfer of a colonial political economy to a domestic one involved a new emphasis on controlling *domes-*

ticity, a new concentration on the political economy of the household. An efficient, well-run harmonious home is a national asset: the quality of the domestic environment has a major influence on the physique and health of the nation. A chain of equivalences is at work here; the prevailing logic runs something like this: If the woman is clean, the family is clean, the nation is clean. If the French woman is dirty, then France is dirty and backward. But France can't be dirty and backward, because that is the role played by the colonies. But there are no more colonies. If Algeria is becoming an independent nation, then France must become a *modern* nation: some distinction between the two must still prevail. France must, so to speak, clean house; reinventing the home is reinventing the nation. And thus, the new 1950s interior: the home as the basis of the nation's welfare; the housewife—manager or administrator and victim, occupying a status roughly equivalent to the *évolué* or educated native in the colonial situation—efficiently caring for children and workers. Or, in a slightly later historical development, the elaborate catalog fantasy of the accoutrements—the new *luxe, calme, et volupté* of modern living proposed in the first chapter of *Les choses.* Here the reader meanders through pages of description of objects and furniture before becoming aware of the existence of a human being, one whose invisible labor of upkeep and maintenance is naturalized, part of the surroundings: "There, life would be easy, simple. All the servitudes, all the problems brought by material existence would find a natural solution. A cleaning woman would come every morning."[13]

Women's magazines played a leading role in disseminating and normalizing the state-led modernization effort. Magazines targeting a specifically female readership were born in France in the 1930s, but they knew a significant surge in number, circulation, and readership in the decade following World War II.[14] *Marie-France* was founded in 1944, *Elle* (with Françoise Giroud and Hélène Lazareff at the helm) a year later; *Femmes d'aujourd'hui* appeared in 1950, and the first reissue of *Marie-Claire,* after a long hiatus during the 1940s, in 1954.[15] The story of the early years of *Elle* is in many ways exemplary; its founder, Hélène Lazareff, had spent five years in the United States working with the best American magazines, including *Harper's*. Among the technological innovations she brought back with her from the States was a perfected use of color unknown in France; she became the first French magazine editor to use color photography. Her colleague at the magazine, Françoise Giroud, describes her: "With her American culture, she was the vehicle for a modernity that, for better or worse, would invade French society. She was made for the world of disposable cigarette lighters, dresses that last for a season, plastic packaging. In a ravaged France, the society of consumption was still far away. But Hélène was already the mouthpiece of its hysteria for change."[16] Together, Giroud and Lazareff constructed the composite portrait of the ideal reader of *Elle;* they called her "the reader from Angoulême" and endowed her with all the frustrations and unmet desires of a war-deprived adolescence. But if the targeted reader was the young woman from Angoulême, the image of femininity constructed by the magazine had more to do with its editor Hélène Lazareff's attachment to what Giroud calls "the *joie de vivre,* the optimism, the generosity emanating from that country [the United States] in those days." The United States was, above all, a "happy country" that possessed "that American health made up of equal portions of optimism and dynamism." And the look projected by the American woman of that time was, for Giroud, one of hygienic self-assurance: "In those days, an American woman was someone whose hair was always freshly washed and combed." The success of *Elle,* she writes, "coincided with the beginning of vast social changes in France, of a break that came out of the war and the lack of consumer goods," and with "the appetite for frivolity, for changes of taste in clothing and dress that the war had wrought."[17]

What Lefebvre called "the domesticated sublime of the world of women's magazines"[18] drew the attention of all the analysts of everyday

Les Américaines
ont les cheveux
les mieux soignés
du monde !

Des millions d'Américaines ont demandé à Helena Rubinstein, le
guide de la beauté, des produits pour avoir de beaux cheveux.
Helena Rubinstein vous offre aujourd'hui ces produits en France.

FIGURE 2.1 Helena Rubenstein advertisement, *Elle*, May 1955.

life in the late 1950s and early 1960s—Morin, Barthes, and Lefebvre himself—each of whom devoted pages of often speculative prose to the phenomenon.[19] An article published in *Esprit* in 1959 by Ménie Grégoire undertook a more systematic analysis, dividing the contents of the four leading magazines into five categories: romance, fashion and beauty, cooking, practical advice, and culture. She then provided a statistical breakdown of the amount of coverage given to each category. Implicit in Grégoire's analysis is the magazines' directive that these five categories and only these categories constitute a woman's life: her schedule, her *emploi du temps*. Women's magazines proposed above all to fill up that schedule, to provide a daily narrative of female existence involving shopping, housekeeping, fashion: daily life is full and complete, and the reader finds a ready-made model of accomplishment, fulfillment, and satisfaction. A frequent use of *sondages* or readers' surveys (emerging for the first time) allowed the magazines to conform to their public, to address, despite regional nuances, *la femme typique,* and, in so doing, to avoid scandal. The French woman of 1959, writes Grégoire, was easily shockable, and shock was to be avoided at all costs.[20] As such, women's magazines were both the result of and the application to the quotidian of a set of techniques oriented by market research.

If these five categories constitute a woman's life, what was left out? First, according to Grégoire, any notion of career ambition: all such interests on the part of women characters in the romantic *feuilletons,* for example, disappear with the advent of passion. Second, formal, even nonpartisan, politics. An article that appeared in *Elle* in 1955 by Françoise Giroud entitled "Apprenez la politique" might appear exceptional; in fact it confirms Grégoire's conclusions. The article is an informal survey that tries to answer the question, "Are French women interested in politics?" Giroud concludes negatively, though she ends the piece by conjuring up an ideal for women to strive for: not to *be* a politician but rather to *be* a woman "who has succeeded in finding her way through the fog and who can in some ways 'follow the game' without herself feeling capable—nor desirous—of playing herself."[21] She should become someone whose political ignorance, in other words, is not advertised. Third, no scientific or economic information is offered to women who in fact buy some 60 percent of the products consumed.

Career, politics, science, and economic information are relegated, in a strict gender division of access, off limits: "In France, politics is a machine, and women detest mechanics."[22] Françoise Giroud's own journalistic career is a case in point. Editor-in-chief of *Elle* until 1952, she is then chosen by Jean-Jacques Servan-Schreiber to codirect with him the new news and information magazine *L'Express*. She in fact runs the magazine, often single-handedly during his long absences in Algeria and elsewhere, for seven years. When she quits in May of 1960, the magazine flounders:

> But what *L'Express* was going to miss the most was the intellectual openness of its female director. Jean-Jacques Servan-Schreiber was only interested in politics. Giroud broadened the horizon of the journal to include literature, the cinema, philosophy, and everyday life. And this void would be felt by Servan-Schreiber more than anyone else:
> "We must find someone to replace Françoise for everything that isn't politics in the magazine. A woman, undoubtedly."[23]

Servan-Schreiber does, in the end, find another woman—another Françoise, in fact: glowing in her success as a novelist and voice of youth, Françoise Sagan replaces Giroud three weeks later to cover "everything that wasn't politics"—she lasts only briefly on the job. But Sagan's first published editorial in the magazine, unexpectedly taking up the case of tortured FLN prisoner Djamila Boupacha,[24] is itself an indicator that the rigid distinction between politics and everyday life, technique and sexuality, men's realms and women's, armies and civilians, is beginning to give way in the France of torture and the Algerian war. Two years later, when the magazine hits another sales slump after the end of the Algerian War, Servan-Schreiber launches a full-scale modernization of its format in an attempt to attract the new, middle-class readership. The magazine "must be powerful as a factory; it must obey industrial laws;"[25] but it must, above all, be produced by offset printing, so that readers will no longer dirty their hands.

According to Grégoire's statistical analysis, it is affairs of the heart—advice to the lovelorn, romantic *feuilletons,* even astrology can be put in this category—that dominates the women's press, with fashion and beauty coming in second. Only one magazine, *Femmes d'aujourd'hui,* specializes primarily in the category that usually comes third: practical (or household) advice [*conseils practiques*]. Under this rubric comes house-keeping, cleaning, household matters, hygiene, and health—in short, that which constitutes the traditional woman's *métier,* then undergoing a dramatic resurgence. "Women are not reluctant to accept help in managing their time and their cares, when it comes to work made up of the most menial, the most monotonous, the most solitary of tasks."[26] Far from being upset, French women greeted with great enthusiasm the arrival, in 1958, of two new magazines devoted to *conseils pratiques* onto what seemed to be an already saturated market: *Femme pratique* and *Madame Express*—the latter published for a brief time in a "conjugal" unit attached to *L'Express,* before divorcing itself off into a separate format. *Femme pratique* eliminated the romantic advice columns and the short stories; it billed itself as "a technical journal for the woman in the home," the "review for the household enterprise." Its first issue was published at 250,000 copies; 450,000 for the second; *Madame Express* knew a similarly spectacular ascendancy. The way for these more specialized organs of household technology was paved by the new kinds of attention given to issues of cleanliness and housekeeping in the standard women's press. The first postwar reissue of *Marie-Claire* in October 1954, for example, declared its intention to "help women feel at ease in the modern age."

That age, it goes on to describe, is "the atomic age but also the age of abundance, of emancipation, of social progress, the age of light, airy houses, of healthy children, of the refrigerator, pasteurized milk, the washing machine, the age of comfort, of quality, and of bargains." The January 10, 1955, issue of *Elle* is devoted entirely to "whiteness": "Beau BLANC, BLANC bébé, boire BLANC"—to the importance, that is, of bleach, of white layettes for babies, of schoolchildren drinking pasteurized milk. One article concentrates on helping women organize their "ideal linen closet": "You have always dreamt of a practical and pretty linen closet in which the family trousseau can be arranged in order. We have realized this dream for you, by choosing new ideas, the best prices, the best quality." An advertisement shows a woman contemplating her own beaming image in a freshly polished stove top: "Et voila! I've finished my stove top. Everything is reflected there!"

But what is reflected there? In a similar image taken from Zola's novel about an earlier stage of commodification in France, *Au bonheur des dames,* a group of women are seen hovering over a pool of colored silk for sale in a department store, entranced: "The women, pale with desire, bent over as if to look at themselves. And before this falling cataract they all remained standing, with the secret fear of being carried away by the irruption of such luxury, and with the irresistible desire to jump in amidst it and be lost."[27] In the roughly one hundred years separating the two images of female narcissistic self-satisfaction, much has changed. In Zola the women see their reflection in a sea of silk and both want and resist the desire to throw themselves in: the relation between woman and commodity is one charged with a surfeit of the eroticism of boundary loss; self-definition via luxury and pleasure (the silk is called "Paris-Bonheur"). In the 1950s advertisement the shining but unyeilding stove-top surface reflects back to the woman the image of accomplishment; there is no give to the surfaces, no tactile dimension, even an imagined one—just smooth shine. The narcissistic satisfaction offered is one of possession and self-possession: clean surfaces and sharp

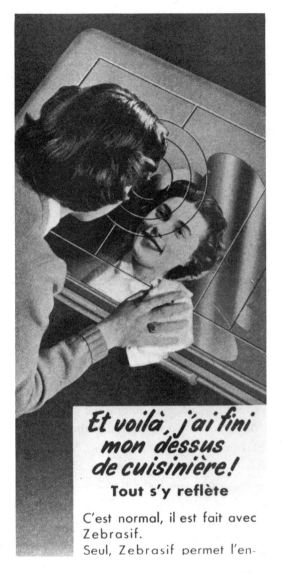

FIGURE 2.2 *Zébrasif* cleanser advertisement, *Marie-Claire*, April 1955.

angles. The completion of a household task completes the woman—everything is reflected there: woman defined midway between the twin poles of domestic science and object fetishism.

The May 1955 issue of *Marie-Claire* contains a how-to guide for "winning the hygiene battle": an article devoted to how best to socialize the next generation of French children to have what the author calls "the cleanliness reflex." The key to making childhood cleanliness an internalized, automatic response is to force the child to repeat a number of ritualized gestures every day; to make children understand the link between dirty hands and illness, for example, make them wear white gloves for an afternoon and at the end of the day show them all the "microbes" on the gloves. "The gesture of taking a clean handkerchief from the cupboard each morning should be as automatic as that of grabbing his notebook for school."[28]

The article concludes with a report on the activities of the "Bureau de la propreté," a subdivision of the Ministry of Education. Beginning in 1953 the bureau sponsored a contest for schoolchildren between the ages of eight and fourteen whose theme was "bodily cleanliness." Perhaps, speculates the author, the success of the contest accounts for the fact that consumption of bath soap increased in the year 1954 by 81 grams a person—reaching 432 grams a year. In fact, items related to health and personal hygiene were among the goods for which demand in France was rising fastest: consumption of these items rose 86 percent in the 1950s.[29] Schoolchildren were not the only competitors in national hygiene forums. A contest for the best housekeeper that originated in the interwar years gained considerable influence in the 1950s. Women were encouraged to compete with each other within the confines of the traditional *métier de la femme* to provide the cleanest and healthiest home environment for their family. To compete at fair advantage meant having the proper accoutrements or tools, as well as the science to make use of them. Claire Duchen has emphasized the importance of domestic science textbooks in this endeavor—volumes such as Paulette Bernège's *De la méthode ménagère,* originally published in 1928 but re-edited for use in domestic science courses in schools throughout the 1950s, devoted to "the rational organization of domestic work, care of the house, sewing and maintaining clothes and linens, whitening and ironing . . . a theoretical and practical teaching of childrearing and of hygiene as well as an initiation into familial psychology and morals."[30] By the early 1950s, however, many Parisian women were also attending the yearly *Salon des arts ménagers* where new and futuristic appliances were displayed in dream arrangements ("a household blessed by God" intones the announcer on the newsreel showing the 1953 *Salon*) and demonstrated by white-gloved technicians.[31] In the 1953 newsreel the viewer is guided through the exhibition by the Dupont couple—statistically the most prevalent French surname—who, we learn, have "located an apartment," that is, triumphed over the postwar housing crisis, and immediately "rushed to the *Salon des arts ménagers* to relearn the art of living well. After years of restaurant eating, they will reacquaint themselves with the recipes of good French cooking. . . . In front of the refrigerators and washing machines Madame Dupont imagined the jealous fit her best friend would have whose entire apartment fit into the Duponts' new bathroom. . . . The apartment of 1953: the ideal place where everything is simple and easy."[32] Or Perec: "There, life would be easy, simple." The newsreel concludes with the Duponts rushing off, their purchases decided, to "the dream they've held close to their hearts for so long": "their first evening 'at home'" ("at home" is in English).

The newsreel narration performs an important function that we might begin to isolate: that of reconciling past and future. Presumably, after having survived the housing crisis that rendered them kitchenless, the Duponts will now *re*learn the traditional French recipes, not in grandmother's dark and somewhat dank kitchen but rather in their new and enviable techno-environment. The authentic *art de bien vivre* is perfectly compatible with the streamlined appliances of modernity; in fact, the appliances of the future, *"les amis de la femme,"* are the best way to

FIGURE 2.3 "The all-electric house," *Marie-Claire,* May 1955.

re-create the meals of the past! One can gain access to the future without loss; nothing is left behind, nothing is wasted.

The brief newsreel narration does much to allay the anxieties of modernization. A wonderful example of Lefebvre's discourse of the "domesticated sublime" ("where one speaks familiarly of the sublime and of the familiar with the tone of the sublime"),[33] the narration brings God into the household and equates dreams and ideals with bathrooms and appliances. Any fears that might arise from having such advanced technological gadgetry *within* the confines of the domestic environment are put to rest: the domestic interior, we are assured, can contain, reenfold the most abrupt technical acceleration. This reenfolding inward—back onto the authentic French life (the old recipes) and the private interiority of the domestic—this reprivatization, is all made possible, of course, by technology, by the unacknowledged opening out onto multinational products and Americanization: the newsreel concludes with "at home" rather than "chez eux."

A young married couple interviewed by Chris Marker in his 1962 documentary about everyday life in Paris, *Le joli mai,* embraces the ideology of a newly privatized domestic life centered on the couple. When questioned if they think about political events—we are on the eve of the signing of the Evian Accords that have brought the eight-year Algerian War to an end—they reply negatively; such things have nothing to do with them, they say; they wish for nothing so much as to "have the pleasure of setting up house" [*avoir le plaisir de préparer son intérieur*]. Marker's own evaluation of the limitations of such a definition of happiness is clear; the footage he includes from a *Salon des arts ménagers* is preceded by the subtitle "The dream is being consumed ready-made" [*Le rêve se consomme tout préparé*]. His linking of the new ideology of the privatized, consuming couple to a national agenda is also explicit: the film ends with the image of a prison in the shape of a hexagon.

This reprivatization of daily life, which we might date as beginning in the early 1950s with the peaking and then slow subsiding of the worst of the postwar housing shortages, modifies and at the same time confirms dailiness in modern France. The task of both modifying and confirming the everyday fell squarely on the shoulders of women. Appliances, after all, were *les amis de la femme;* they formed new links between the woman and the society that created them; they imposed a new set of comportments and behaviors. The woman in the *Marie-Claire* advertisement sees reflected in her gleaming stovetop a new criterion for fulfillment and satisfaction, a new identity; Madame Dupont sees reflected in the washing machine she is purchasing the jealous fit of her friend. The commodity form does not merely symbolize the social relations of modernity, it is the central source of their origin—in this case, a new arena of competition between women. But the ultimate competition for the French woman, her distant horizon of excellence, was the American woman who washed her hair every day. Modernity was measured against American standards; materials imported from America—stainless steel, Formica, and plastic—were valued both for their connotation of modernity and because they were easy to clean.

Subsequent newsreel clips of the *Salon des arts ménagers* abandon the Duponts as guides and offer instead a pair of women to take the viewer through the show. The women presumably constitute friend/competitors rather than a domestic unit; their fictional husbands, we might also presume, are off attending the equally popular yearly event, the *Salon de l'automobile.* In Rochefort's *Les petits enfants du siècle* husband and wife battle over which large-scale commodity to buy next when the government childbirth allocations come through: a car for the husband or a refrigerator for the wife. It is left to the teenage narrator to attempt to find noncommodified affection and pleasure in the arms of an Italian immigrant construction worker. (As in *Les stances à Sophie* the male Italian peasant offers a surfeit of masculine directness in a world otherwise hopelessly mediated by cumbersome consumer durables.) Whereas Rochefort and Simone de Beauvoir focused on the automobile as metonymy for the masculine embrace of the values and privileges of technocracy, two somewhat earlier texts, Elsa Triolet's 1959 *Roses à crédit* (the opening volume of her fictional trilogy, *The Age of Nylon*) and Boris Vian's song from the late 1950s, *Complainte du progrès,* feature men as the gender immune to new, modern desires, the untainted repository for some older, precommodity form of romance or desire—a nonreified desire. (The Situationists also believed in "love" or "desire" as an alternative to commodification; their wry comment, "Given the choice between fulfilling love and a washing machine, young people in the U.S. and the Soviet Union both choose the washing machine," still holds out the possibility of making the other choice, that is, the belief in a desire that is not determined by capitalism, a "pure" desire outside reification.) Triolet's novel, which brought her her first commercial success since she became the first woman to win the Prix Goncourt in 1945, is essentially a morality tale about a young provincial girl, transfixed by the gleam of commodities, who succumbs to newly available credit to outfit her new "modern" existence in the city, who struggles with skyrocketing debt, and who finally returns full circle to her abject origins to die a disfiguring and wretched death at the hands of sharp-toothed natural elements. The opening chapters of the novel, which recount the young Martine's acquisition of a subjectivity and an ambition sufficient to propel her from the depths of the virtually medieval forest where she lives with her countless siblings and promiscuous mother (whose sheets, washed only twice a year, emit foul odors that repel the child), attribute that acquisition to a finely developed relationship to hygiene: "She didn't know why dirty sheets, snot, rats and excrement from time to time made her vomit."[34] The tale that follows is in many ways an ur-narrative of what Edgar Morin (who grants the exact same history a positive, liberational valence) calls the "decolonization of the peasant woman."[35] For Morin too the acquisition of a "filth complex" is the first step in a process of gaining psychological autonomy and an expanded personality and horizon that will eventually propel the peasant woman out of the countryside. The rural woman's construction of a cleanliness threshold is likened by Morin to the psychological process analyzed by Fanon: the formation of a new subjectivity on the part of colonized men engaged together in violent revolutionary struggle. Morin charts the formation of the new subjectivity as follows: first the media infiltrations—via press or radio—that bring advertisements of new ways of comfort and physical hygiene to the depths of the countryside; then a hard-won acquisition for the home (farmers, if they were inclined to modernize, spent all of their money on the outdoors—on tractors, specifically, and resisted purchases for the home): some initial commodity that in turn creates new motivations for comfort. The crucial stage is reached when what Morin calls "the crystallization of the interior" occurs: a kind of drawing of a definitive boundary between interior and exterior that gives the wife a "realm of her own," and, by extension, a new psychological interiority and depth that will lead to autonomy. The psychological liberation is dependent on the Manichean division of the interior from the exterior, the latter now viewed as sordid, dirty, and repellent to the woman, who vigilantly polices its various invasions into her realm in the form of grime and odors tracked in from the outside on bodies and hands, even under fingernails. (Morin quotes a fifty-five-year-old farmer speaking of the women in his village: "They don't like dirt under their nails; they want to put red polish *on* their nails.") All of the new repulsions and aversions coalesce into a "filth complex" that accompanies the definitive eruption of the new domestic model into the female psyche and which in turn translates into a global repudiation for the peasant condition.

Through reconquering and modernizing the domestic interior, rural women accomplish their own psychological modernization that will eventually propel them in a near-unanimous feminine migration out of the servitudes of the countryside to urban areas, themselves associated unequivocally with liberation. Morin sums up the dialectic as follows: by closing the farmhouse off to the earth, you open it up to the world.

Morin's decolonization narrative—he refers to French agricultural laborers as "the wretched of the earth"—was based on an ethnographic study he performed in the mid-1960s of a village in Brittany called Plodémet. Triolet's character Martine is born in a village that is only sixty kilometers from Paris but whose fundamental rhythms, it seems, have gone unchanged since the Middle Ages; her desires and her own flight to Paris are represented at least initially as no different from those of the women Morin encountered in Brittany. Like them, her subjectivity is formed through the "crystallization of the interior"; but because she is just a child and her own house is irredeemably filthy, she must find it elsewhere: in the clean, shiny universe of feminine beauty and hygiene, the hairdresser's shop. Her ascendance into this adopted world is given all the importance of a religious conversion: "No palace out of *A Thousand and One Nights* could have overwhelmed a human being more, all the perfumes of Arabia could never have given anyone the intense pleasure Martine felt in that little house saturated with the odors of shampoo, lotions and cologne" (p. 39). The chapter that recounts her first bath is entitled "The Baptismal Font of Modern Comfort" and, as an evocation of the domesticated sublime, is worth quoting from at some length:

> When Martine saw the bathtub for the first time and Cécile told her to soak herself in all that water, she was overcome by an emotion that had something sacred about it, as though she were about to be baptized. . . . Modern comfort happened to her all in one fell swoop, with running water, gas heating, electricity. . . . She never became completely used to it when Mama Donzert said to her: "Go take your bath" . . . she felt a delicious little thrill. . . . The tile of the bathtub was smooth, smooth, the water was gentle, gentle, the bar of soap, all new, produced pearly suds . . . a pink and sky-blue sponge. . . . The milky light-bulb lit up every innermost

recess of the bathroom, and Martine scrubbed every inner-most recess of her body with soap, pumice-stone, brushes, sponges, scissors. (pp. 39–40)

It is in this way that Martine begins her transition "from one universe to another" (p. 47).[36]

But whatever sacred, purifying, or expansive properties that are associated with her newly acquired rituals of beauty and hygiene degenerate over time after she moves to the city. Over the course of the novel they are reduced to a set of obsessional character traits and compulsions that control Martine's life and, far from expanding her world as Morin would have it, actually limit her actions and destroy what affective life remains for her. Martine's husband is powerless to influence her and is progressively excluded from her life: "How could he compete with Martine's ideal press-button universe? She was a savage dazzled by the brilliant baubles the white men dangled in front of her. She adored modern comfort like a pagan, and she had been given credit, the magical brass ring from the fairy-tales that you rub to make a genie appear to grant your every wish" (p. 197).

Both Triolet and Morin draw an analogy from the colonies to describe the situation of French rural women. But where Morin sees the move to the city and the resulting change in social consciousness as a "decolonization" of "the wretched of the earth," Triolet shows her savage to be a dupe of empty promises, blinded by the shiny surfaces of modern appliances, enslaved to an endless spiral of debt, and if anything, newly and more inextricably colonized. Martine's class and regional origins, to which she must ineluctably return (having repudiated them so completely) act to exclude her from the gains and pleasures of modernization.

But the unhappy fate of female characters as different from Martine as the four shopgirls working in an appliance store selling *electro-ménagers* in Chabrol's brutally realist 1960 film *Les bonnes femmes,* or Elise in *Elise ou la vraie vie* newly arrived from the provinces to work in the car factories, suggests that the difficulty they all share has more to do with what Adrian Rifkin has called the difficult conjugation of women with urban pleasure. "What is pleasure for women as a subject in the city, and why is it not the same as man's? Why is one's libertine and the other's 'normal'?"[37] In *Elise ou la vraie vie* Elise's overdetermined childhood search for a room—a self, interiority, an identity separate from her brother's—is transformed into her late-night treks across Paris with her Algerian would-be lover Arezki; multiple layers of surveillance, from the factory to the police, from the FLN to even her leftist pro-Algerian brother, prevent them from finding a room in which to be together. Her brother, on the other hand, a provincial and a worker like herself, has rooms and lovers to spare. In Chabrol's film, what Edgar Morin calls "the surprising conjunction between feminine eroticism and modern capitalism, that looks for ways to stimulate consumption"[38] is made all too clear, as the camera looks in the shop windows to show the girls arrayed, displayed, and waiting, posed next to the washing machines and vacuum cleaners, in a shop situated tellingly next door to the Grisbi striptease joint. Unlike Martine, these girls want love, not a dishwasher. Adrift in romantic fantasy fed by the urban setting, they suffer the unchanging ennui of the workplace, the petty tyrannies of the shop owner, and a bare, celibate, quasi-dormitory home life not unlike that of immigrant male factory workers. The frustration (verging on tragic consequences) of the shop girls is juxtaposed to the traditionally easy, confident to the point of sadistic, access of the men in the film—from the ageing shop owner to the frequenters of the Grisbi Club next door—to a variety of urban erotic pleasures.

Boris Vian's song "Complainte du progrès," a kind of contemporary *Au bonheur des dames,* shows women captivated by the immediacy of the commodity world being transformed into commodities themselves. His complaint against progress echoes that of Martine's husband, powerless to compete with the gadgets women want. Men too then, along with women, the elderly, the young, the working class, the traditional petite bourgeoisie, immigrants, the peasantry, intellectuals, and other cultural elites, could construct themselves as the primary victims of capitalist modernization:

> Autrefois pour faire sa cour
> On parlait d'amour
> Pour mieux prouver son ardeur
> On offrait son coeur
> Aujourd'hui c'est plus pareil
> Ça change, ça change
> Pour séduire le cher ange
> On lui glisse à l'oreille
> Ah . . . gudule . . . viens m'embrasser . . .
> Et je te donnerai
> Un frigidaire
> Un joli scooter
> Un atomizer
> Et du Dunlopillo
> Une cuisinière
> Avec un four en verre
> Des tas de couverts
> Et des pelles à gateaux
> Une tourniquette
> Pour faire la vinaigrette
> Un bel aerateur
> Pour bouffer les odeurs
> Des draps qui chauffent
> Un pistolet à gaufres
> Un avion pour deux
> Et nous serons heureux.[39]

[Before when you went wooing/You spoke of love/To prove your passion/You offered your heart/Today it's not the same/ It's changed, it's changed/To seduce your dear angel/You whisper in her ear/Ah . . . darling, come kiss me/And I will give you/A frigidaire/A shiny scooter/An atomizer/And some Dunlopillo/A stove/With a glass oven/A pile of covers/ And cake pans/An eggbeater/To make vinaigrette/A beautiful odorizer/To eat up odors/Warm sheets/A waffle iron/An airplane for two/And we'll be happy]

Cars and refrigerators as objects of choice to designate the sexes; the iconography is most apparent in film, where directors made use of the sheer size of these cargo-cult objects in shocking visuals. Unlike television, subsequent audiovisual information, or computer technologies, the car and the refrigerator have the iconographic advantage of having a single physical unity; each is a "total object" in and of itself. Perhaps the most memorable scene in Dino Risi's *Il sorpasso* is one that verges on a kind of gender aggression. Hot-rodder Vittorio Gassman drives his sports car so recklessly on a narrow and remote country road that he causes a truck to overturn, spilling its load of shiny new white refrigerators onto the dirt; the camera lingers just a second on the incongruous refrigerators, glistening like so many beached whales in the sunlight. Jacques Rozier's *Adieu Philippine* uses a similar economy to designate the separate realms of the genders: Michel, a young technician who has just co-purchased (with three male friends) his first car, befriends two women who act in TV commercials. In the first commercial audition the two women are dressed alike and surrounded by thousands of identical boxes of laundry soap; hundreds of takes fail. Their breakthrough happens when they sign a contract with a refrigerator firm: dressed as Eskimos and surrounded by penguins and igloos on a fake ice floe, they recite the slogan, "Even in the North Pole, you need a refrigerator." But perhaps the most heavily weighted use of the gendered iconography is found in Jacques Demy's perversely realistic historical musical about an automobile mechanic who gets drafted, *Les parapluies de Cherbourg*. In this film the length of time that transpires between the mechanic's departure for Algeria and his return is measured not by some representation of the hardships of war (unseen in the film), nor by his lover's (played by Catherine Deneuve) pregnancy and childbirth, nor by her abandoning him for another man whom she subsequently marries, but rather by the transformation—seen through the dismayed eyes of the returning soldier—of the quaint umbrella boutique where Catherine Deneuve and her mother once struggled to make ends meet and where the lovers had spent so many happy afternoons, into a cold-looking store that sold washing machines and refrigerators.[40]

The object itself, the refrigerator or "cold spot" as it was called in its early incarnations in the United States, with its pressed steel casing and seamless finish, conveyed the image of absolute cleanliness and newfound hygiene: its brilliant white finish was the physical embodiment of health and purity. The refrigerator as mass object of desire and one of the "mature" consumer durables was indeed the object-fetish for the new modernized home;[41] its arrival into Martine's modern Parisian apartment in Triolet's *Roses à crédit* is granted its own ironic paragraph: "The frigidaire had appeared in the kitchen in the middle of the winter. It was enthroned there like Mont Blanc, handsome, cumbersome and useful" (p. 159). But the object, though fetishized, was less important in and of itself than in its contribution to a total environment, its efficient "communication" with other appliances. Because of the need to provide a physical and emotional separation from the place of work, the nineteenth-century middle-class French home was organized around the sitting or drawing room, the soft, textured, plushy slipcovers and casings of which were so central to Walter Benjamin's account of the emergence of the detective story during the Second Empire.[42] Gradually in the twentieth century, and certainly by the 1950s and 1960s, the modernized kitchen became the focal point of family life, as countless

FIGURE 2.4 *Les parapluies de Cherbourg*

FIGURE 2.5 *Frigéco* advertisement, *Elle,* May 1955.

FIGURE 2.6 Brandt refrigerator advertisement, *Marie-Claire,*
May 1955.

fictional and filmic representations made clear: "They always ate break-fast together in the kitchen, on a light green table made of one of those brilliant materials that was always clean. Coffee in a beautiful, perfected coffeemaker, butter, jam, toast . . . flowered bowls and stainless silver-ware."[43] The kitchen was also the centerpiece of a rationalized home no longer concerned with differentiating itself from the workplace. Claire Duchen has shown how the directives issuing forth from home-eco-nomic textbooks and women's magazines about the management of the home unabashedly adopted a Taylorist organization program involving a clear distinction between the direction and the execution of tasks and the organization of both space and time to increase production. House-wives were encouraged to perform (or to have performed for them) labor-saving analyses that would help them reduce unnecessary effort and movement and eliminate "useless gestures." Whole articles were devoted to the arrangement of appliances in the ideal kitchen or laundry room: the housewife should be able to proceed from one to another in assembly-line fashion without retracing her footsteps. "Work should advance through space in a continuous, straight line, without useless to and fro or going backwards. Study the entrances and exits, the juxta-position of rooms to each other in order to make possible this advance-ment of work in a straight line."[44] For Duchen these developments had the effect of simultaneously elevating the woman and infantilizing her: on the one hand, her Sisyphean task was a science, requiring logical expertise, but on the other, she was newly dependent on authorities outside the home. No longer was a commonsense response, or the vague memory of how one's grandmother performed a task, sufficient—ex-perts must be consulted, precise timetables kept to. By reducing the difference between the home environment and the workplace, house-work began to look more like real work—but not to worry, for the domestic appliances, essentially middle-class replacements for the nine-teenth-century servants, the new *amis de la femme,* were there to lighten the load. Adrian Forty has shown how manufacturers followed the

FIGURE 2.7 Hoover advertisement, *Elle,* October 1954.

analogies between home and factory and styled their appliances in forms reminiscent of factory or industrial equipment in order to emphasize the labor-saving efficiency they were claiming for their product.[45]

Was the housewife an assembly-line worker, then, or a white-collar manager, issuing orders to an army of worker-appliances? Certainly the discourse of the rationalized household in Bernège, and in the countless advertisements showing a beautifully dressed and bejeweled woman vacuuming in her high heels, worked to promote the latter position. But her ambiguous status placed the housewife in an analogous situation to that of the *jeune cadre* in the factory structure, elevated above the immediate travails of the assembly line but no less governed by the time clock. And Forty is among the many who dispute the idea that time was actually saved by housewives with the introduction of "labor-saving" devices, citing evidence that domestic appliances cause more time to be spent on housework and not less. Labor-saving appliances save no labor, if only because their introduction into the household is accompanied by a rise in the standards and norms of cleanliness. Nor does the repetitiveness of the tasks change; Christiane Rochefort represents her character Céline in *Les stances à Sophie* (a woman with a fully modernized house *and* a Spanish maid) as brought up short by the realization that, in her capacity as steward [*intendant*] of the household, she will have to ponder the question of what its members are to eat for lunch every day of the year for the rest of her life: "And look at the amount of time wasted on that insane activity, one that has to be done again the next day, and the day after, and every day, and to think that there are 365 of them in one year alone and that we don't know how many years there are, and that on every one of those days the question will be asked and must receive an answer."[46] Even the mental work of housework exhibits the fragmentation and repetition characteristic of manual labor. And the new level of women's dependence introduced by the modernized home—on husbands, to buy the appliances; on the opinions and directives of experts and specialists, to run them and organize them—suggests that the real decision-making power, the *savoir-faire,* has shifted outside of the woman's immediate sphere of control.

KEEPING HOUSE

Writing in 1966, Jean Baudrillard notes a decisive historical change surrounding this idea of the home and the labor taken to maintain it: "This is no longer the traditional housekeeper's obsession: that everything be in its place and that everything be clean. The old one was moral, today's is functional . . . everything must communicate with everything else."[47]

"Everything communicates." In Jacques Tati's treatise on the anxieties of modernization, *Mon oncle,* the obsessively clean housewife character, Mme. Arpel, repeats this line over and over whenever she shows off her home to guests. "Everything communicates": the line proudly sums up a space designed to promote efficiency of movement and flow of bodies from one room to another, a kind of interior circulation or traffic like the one we see automobiles involved in outdoors. The joke, of course, is that communication is exactly what is lacking in this sterile, precise, fenced-in suburban home where parents relate to their sullen, silent child in a series of compulsive directives about hygiene: don't mess up your room, put your books away, wash your hands, hang up your clothes, and so forth.

Baudrillard's observations about "functional" cleanliness replacing an older, premodern or moral cleanliness find an echo in a remark made by Barthes in the early 1960s concerning the new French vocabulary springing up to denote the desired "shine" of an automobile. "We want it [the car] to be more than clean: *bichonnée* [caressed, pampered, 'cherried'], *briquée* [scrubbed], *lustrée* [waxed and shined]." The desire for the car's shine, according to Barthes, is the desire "to remake the virginity of the object over and over again, to give it the immobility of a material on which time has no effect (the obsession with cleanliness is certainly a practice of immobilizing time)."[48] In Barthes's description as well, the moral value of cleanliness is superseded by a more pressing imperative; to make virgin here is less a moral activity than one that involves making something absolutely (and eternally) new: the object outside history, untouched by time—the object and one's relation to it unchanging, as in a functionalist equilibrium, endlessly reproducing itself.

The desire to immobilize time: five years after commenting upon the "great hunger [*fringale*] for cleanliness" sweeping France, Barthes goes one step further, linking the will to cleanliness here with a desire to immobilize time, to step outside of history, or perhaps, by extension, to retreat inside a controlled, rationally created environment superior to one engendered historically. This movement of retreat, or *repliement* ("folding back inward")—the dominant social movement of the period— was theorized at the time by Lefebvre and Cornelius Castoriadis under the name of "privatization." Privatization is certainly nothing new; its historical particularity or palpability in the late fifties and early sixties can only be seen as the result of an *acceleration* in the process by which various spheres of life become progressively separated from each other— the most crucial being that of domestic life from the sphere of work. For Castoriadis privatization constitutes the most striking trait of modern capitalist societies, if only because it is not particular to the working class but is found among all social categories. It emerges when a society's most important characteristic becomes its success in destroying the political socialization of individuals, such that one experiences public or even social matters not only as hostile or foreign but also as out of one's grasp, unlikely to be affected by one's actions. People are thus sent more firmly back into a retrenchment in private life, from which they attempt to fashion an anchor of sorts, particularly since the value of work too has undergone increasing decomposition with the growth of bureaucracy. In the end, privatization for Castoriadis is a manifestation of "the agony of social and political institutions that, having rejected the population, are now rejected by it."[49]

With the decline or decomposition of identities based on work or social collectivities, what remains? For Lefebvre the answer lies in the qualitatively new way that everyday life, private or "reprivatized" life, and family life—or, as I would put it, the idealization of the couple— are intimately linked in this period around the identity of the home dweller, the *inhabitant,* and the practices associated with the sole remaining value, *the* private value par excellence, that of consumption. To be "at home," like the Duponts in the newsreel, like the married couple in *Les choses,* like the couple interviewed in Marker's documentary, is to have an identity, one based on security and permanence that state-produced anxiety and the state-produced compensation for that anxiety have gone a long way in helping create. "The role played by attachment to métier is played by attachment to space,"[50] sums up Alain Touraine in *La société post-industrielle.* Lefebvre is less abstract in this instance, his remarks informed by the numerous contemporary empirical studies of the most widespread French middle-class and working-class fantasy of the period: the house-in-the-suburbs [*pavillon*] fantasy.[51] Attachment to space in this instance is not some regionalist sentimentality but rather something quite specific: the desire to be a homeowner.

> The owned or co-owned dwelling (half of French people own their dwellings), the vacation house, not only serves an economic function but security-providing (and thus identity-providing) function as well. Their purchase constitutes a placement. . . . The owner of a single-family dwelling [*pavillon*]. . . . is there for life. He has his place in space. He prolongs himself in the Same without the Other taking it away. He is established in the identical, the repetitive, the equivalent. The enduring nature of goods both symbolizes and realizes the permanence of an ego. That ego certainly lives better in its own property than in an anxious state within a lodging it could lose from one day to the next. These

trivialities make up the triviality and, thus, the force of the quotidien.[52]

Privatization, or losing oneself in the repetitions and routine of "keeping house," meant an increasing density in individual use of commodities and a notable impoverishment of interpersonal relations. For both Lefebvre and Castoriadis it constituted above all a flight from history. This flight was not to be construed as the absence of history but rather itself a historical symptom founded on the wish to make the world futureless and at that price to buy security.

Laurence, the main character in Simone de Beauvoir's *Les belles images,* is, like Tati's Mme. Arpel, the embodiment of the new privatized French middle-class woman, the inhabitant: tugging at her consciousness throughout the novel is the glimmer of a wider world, more vital, perhaps, than her preoccupations with her troubled daughter, her dissatisfied lover, her desk drawers, and the intrigue at the advertising agency where she works. Several times she attempts to finish reading an article she has begun in a magazine concerning the ongoing torture in Algeria; inevitably her thoughts turn to shampoo.

Torture, shampoo. A contemporary cartoon by Bosc reiterates Beauvoir's metonymy, making the relation between the new, modernized hygienic France and the *"sale guerre"* across the sea more explicit.[53] The cartoon shows a French paratrooper in camouflage bending over a sudsy bathtub, his hands submerged; a box of Pax laundry soap ("extraordinaire pour la lessive!") stands next to the tub—but a man's feet stick out of the water; the bubbles aren't suds, but rather the tortured man exhaling.

To put a *para,* as they were called, in the place of a housekeeper is to show in what sense Algeria, far from constituting the "other" of France in this period, is better seen as its monstrous and distorted double. Double in that Algeria, like France, will be the scene of some violent housecleaning; distorted, in that French *men,* who would never lift a

Figure 2.8 Bosc, "Il faut que la torture soit propre." Courtesy Musée d'histoire contemporaine—BDIC.

finger to do housework at home in France, are put to work in the homes of Algerians. "In Algeria, the proletariat is more proletarianized than in metropolitan France, and the bourgeoisie more bourgeois, the petit bourgeois more acerbic and the feudalisms more feudal. And the French Army is more Army."[54] Nowhere was this "doubling" effect more apparent than in the actual conduct of the war. For in revolutionary Algeria of the late 1950s and early 1960s, just as in France, the category of the "inhabitant" was assuming a new priority—at least in the writings and practice of the principal French theorist of "psychological action" (torture) and "pacification," Roger Trinquier. In his influential 1961 book, *La guerre moderne* (immediately translated into an American edition), Trinquier argues that the newness of this kind of war could be attributed to two dimensions: first, to the extent of its actions, which must include the political, economic, psychological, military, etc.; and second, to the lack of definition of the enemy. This last problem is a result, in part, of the altered spatial dimensions of this type of war—the lack, that is, of a delineated battlefield:

> Military schools teaching classic doctrines of warfare rely upon a number of decision factors—the mission, the enemy, the terrain, and the resources.
>
> But one factor that is essential to the conduct of modern warfare is omitted—*the inhabitant.*
>
> The battlefield today is no longer restricted. It is limitless; it can encompass entire nations. The inhabitant in his home, is the center of the conflict.[55]

For the Algerian, to be "at home," an inhabitant, is to be at the center of the conflict: the Algerian inhabitant, unlike the French, is not necessarily depoliticized or privatized; to be at home is possibly the most politicized state, the most connected, in solidarity with, the most vitally a part of, the nationalist struggle. In Algeria during these years, as in

France, the *inhabitant* is central, the status of home dweller newly important, but the identity is inverted.

So too do the newly modernized French interiors and techniques, the electricity and indoor plumbing, appear in a distorted, nightmarish guise in their narrative reflection across the sea. To the extent that information could bypass the heavy state system of seizures and censorship activated at the time to condition or cleanse the image of reality reaching French citizens of the international tensions they were involved in, everyday words and everyday places—kitchens and bathrooms—were beginning to take on new and horrific dimensions.[56] The dark underside of French comfort emerges most clearly in the settings described in the first (and most widely read) of the personal accounts of torture at the hands of the French military to be seized by the government, that of French communist Henri Alleg; in fact, the legitimacy of Alleg's testimony was later ascertained by the military authorities because of his ability to describe by memory several of the rooms in the torture building of El-Biar, but particularly the kitchen, where he would never have been taken if the interrogation had been conducted normally.[57] El-Biar, as it emerges in sidelong glimpses throughout Alleg's narrative, is actually a large apartment building under construction: "The bars in the reinforced concrete stuck out here and there from the masonry; the staircase had no balustrade; from the ceilings hung the wires of an unfinished and hasty electrical installation."[58] Alleg duly notes the sparse furnishings of the various rooms where he is taken, whose existence in a half-built nebulous state of potentiality seems to conjure up the inhabitants of the future, recalling in a grotesque inversion the opening chapter of Perec's *Les choses,* where the couples' "dream apartment" is laid out meticulously in the conditional tense:

> Your eye, first of all, would glide over the grey fitted carpet
> in the narrow, long and high-ceilinged corridor. Its walls
> would be cupboards, in light coloured wood, with fittings

of gleaming brass. Three prints, depicting, respectively, the Derby winner Thunderbird, a paddle-streamer named Ville-de-Montereau, and a Stephenson locomotive, would lead to a leather curtain hanging on thick, black, grainy wooden rings which would slide back at the merest touch.[59]

Like the reader of Perec's opening chapter, the reader of Alleg's testimonio is eventually, in the course of Alleg's one-month stay at El-Biar, taken on a full tour, shown the whole house:

> I was taken into a large room on the third or fourth floor behind C . . . , apparently the living room of a future apartment. Several collapsible tables, blurred photos of wanted suspects on the wall, a field telephone: these made up all the furniture." (p. 47)
>
> One floor down, I entered a small room on the left of the corridor, the kitchen of the future apartment. There was a sink and an earthenware cooking stove, surmounted by a shelf on which the tiles had not yet been laid and only the metal frame was in place. (pp. 48–49)
>
> At the end of the corridor I was taken into a cell on the left-hand side; this was in fact a bathroom without fixtures. (p. 85)

In this "kitchen of the future" the water tortures and trial by fire inflicted on Alleg's body by French paratroopers become parodies of domestic functions. Alleg is electrocuted by telephone (the *magneto* was a telephone turned into a torture device) and drowned at the kitchen sink. Everyday objects of comfort are turned against him: "I tried to lie on my stomach but the mattress was stuffed with barbed wire" (p. 66). In narratives from the wars of decolonization, familiar objects appearing in a routine inventory can become metonymically ominous through their proximity to torture implements, as in this matter-of-fact description of his field tent by a French officer in Indochina: "Here's my desk, my table, my typewriter, my washbasin, and over there, in the corner, my machine for making people talk. The dynamo, I mean."[60]

Or, as in the case of Djamila Boupacha, raped by French soldiers using a toothbrush and a bottle, any houshold object can change functions, be altered. Reading the following inventory from a narrative by an Algerian student tortured in Paris, it is difficult to know what to fear most: "I was led into a room, where they took off my blindfold and handcuffs. I saw two benches about six feet long; two wooden tables; a basin about twenty inches in diameter, full of dirty water, some empty champagne bottles whose necks were stained with blood; a piece of soap; a pile of ropes and rags."[61] The *gégène*—"a genuine product of civilization"[62]—was a simple electrical apparatus, easily accessible or simple to construct, consisting of wires attached to an electric plug. Any do-it-yourselfer could handle it. And this is exactly how some of the early testimony from draftees about their activities in Algeria (collected primarily by Catholic magazines) sounds—like that of *bricoleurs,* misguided weekend amateurs, clown repairmen: "I myself participated by hooking up electricity to the plastic tubing, and to the bathtub."[63]

The new techniques that were revolutionizing the countryside in France—the arrival, that is, of bathtubs (running water) and electricity[64]—were putting a modern, more hygienic touch on torture as well: "Before and after the Algerian War, torturers boasted about employing a *clean* torture, one that didn't leave any traces, and, to the extent that there was 'progress' in the techniques of repression, it's obviously there that it took place: the use of electricity leaves fewer traces than pulling out a tooth or a fingernail."[65]

The striking prevalence of the ideologeme of cleanliness in the writings of the period and the struggle over who—what group, what institution, what race, what generation, what gender—was to claim it as a constituent quality suggests that more is at stake here than the wish to dispel or deny "the horrors of war" in general. World Wars I and II, unlike the Algerian War, were "clean" wars; the distance separating the purity of those conflicts from the current one provided one of the major chasms dividing the French working class during those years: older French workers who had served in the First or Second World War were unsympathetic to the young generations of workers, not considering their war to be "a real war."[66] A Catholic draftee character in Maurienne's 1960 fictionalized testimonio about draft desertion, *Le déserteur,* puts it this way: "I have a father and a grandfather who fought in wars," said Alain, "but they did it more or less cleanly [*proprement*]."[67] The Algerian conflict, from its earliest moments, was known in France as the *sale de guerre*[68]—that is, when its status as a war was even acknowledged. In the final moments of the French empire, the old colonial rhetoric of the "civilizing mission" must be revved up, bolstered, in order to justify what was beginning to leak through the cosmetic discursive blanket thrown over Algerian affairs. In other words, at the peak of the empire's most barbarous behavior, the barbarity of the *Algerian*—his need of cleansing, schooling, civilizing—must be all the more certified.[69] Thus, various specific military operations are given "cleansing" names and functions. In the opening pages of his fictionalized testimonio, *Lieutenant en Algérie,* Jean-Jacques Servan-Schreiber speaks of the process known as "cleaning up a casbah;"[70] in army parlance, the nightly patrols by a hundred French tanks along the Morice Line, the sweeping back and forth the length of the two-hundred-mile border to deter border crossings by the FLN in and out of Tunisia, were known as "floor polishing."[71] Perhaps the most striking narrative of internalization of a clean/dirty, civilized, barbarous justification occurs in the testimony of one French draftee:

> They used to ask for volunteers to finish off the guys who had been tortured (so that there would be no trace of it and no danger of stories later). I didn't like the idea—you know

how it is—shooting a guy a hundred yards off in battle— that's nothing, because the guy's some way off, and you can hardly see him. And anyway he's armed and can either shoot back or buzz off. But finishing off a defenseless guy just like that—No! Anyway I never volunteered and so in the end I was the only one in the whole sector who had never finished off "his" guy. I was called "chicken" ["*p'tit fille*"] One day the captain called me out and said: "I don't like having chick- ens around—get on with it, the next one's yours." Well, a few days later, there were eight prisoners who had been tortured and who had to be finished off. They called me and in front of all the boys they said: "He's yours, chicken, get on with it." I went up to the guy. He looked at me. I can see his eyes looking at me now. The whole thing revolted me. I fired. The other chaps finished off the rest. After that it wasn't so bad, but the first time—I tell you that turned me up. It's maybe not the cleanest job, but when you think about it, all those guys are criminals really, and if you let them live they'd only go on killing old men, and women and children. You can't let them carry on like that. So really we're cleaning up the country [*on nettoie le pays*], ridding it of all the scum [*racaille*].[72]

On the Algerian side the struggle to be "clean" is no less pro- nounced. In his afterword to Alleg's *La question,* Sartre argues that what is being fought out in the torture chamber is the whole question of *species:* only one of us is a man.[73] In other words, at precisely the moment that the colonized demand, through their solidarity within a collectivity, the full status of "human," they must be made to designate themselves as humiliated, broken, less than human, animals. Roger Trinquier also admits that information gathering is secondary to the true purpose of torture. For him torture was not merely a way of obtaining information at any price but, more important, a means of destroying, in each fallen individual, the sense of solidarity with an organization and a collectivity. That this was the desired effect can be seen in the testimony of an Algerian student tortured in Paris who seeks to prevent this loss of the collective; for him, staying "clean" is equivalent to remaining part of the whole: "In the middle of the worst tortures, I thought hard of my brothers and sisters, of Ben M'Hidi, of Djamila; and I constantly repeated to myself that one can be covered with filth [*immondices*] and yet remain clean."[74]

Similarly, the Algerian workers on the Citroen assembly line in Claire Etcherelli's *Elise ou la vraie vie* are represented as shunning the soiled *bleu de travail* worn by French workers in favor of clean street clothes: turtlenecks and tweed jackets; Arezki, an FLN organizer, buys an expensive snow-white dress shirt from a shop on the Boulevard Saint Michel simply because, as he puts it, no one would expect an Algerian to have such a thing. The equation of cleanliness with dignity-under-attack is such that in this novel the tactic spreads to the few white French women working in the factory:

> They arrived in the morning, faces made up and hair arranged, and somehow managed during the day to retire and put on fresh lipstick. There was something there that went beyond coquetry: a display, an instinctive defense against a kind of work that ended by reducing you to the level of a bum—the nail polish more often than not hid the grime; the dirty hair was beribboned with velvet; they patted the gray sweat with powder. I can still see my neighbor in the coat-room, a woman of thirty-five, not pretty, wrinkled, forced by the regulations to wear a discolored denim uniform, and who, while driving a Fenwick, kept on her pumps.[75]

FIGURE 2.9 *Elise ou la vraie vie*

For Fanon, for whom torture was simply inherent in, the very logic of, the colonialist configuration, it was imperative that the Algerians, on their side, fighting their war of liberation, must "do so cleanly," "without barbarity": "The underdeveloped nation that practices torture thereby confirms its nature, plays the role of an underdeveloped people. If it does not wish to be morally condemned by the 'Western nations,' an underdeveloped nation is obliged to practice fair play, even while its adversary ventures, with a clear conscience, into the unlimited exploration of new means of terror."[76]

Toward the end of his stay in El-Biar, Henri Alleg reports feeling heartened that the marks and scars from the torture sessions he has undergone are still visible on his body; he takes this as a sign that he will not be executed: "If they had decided to execute me, they had to have (other than the normal bullet wounds) a 'clean' body in case of autopsy" (pp. 105–106). A clean, unmarked body—or, in the words of the draftee I quoted earlier, "so that there would be no trace of it and no danger of stories later." As the war progresses, the moral sense of trace (the mark of guilt, of wrongdoing) gives way to another: traces must be eliminated that mar the clean functioning of an immense system, that throw a wrench in the works, that force a temporary shutdown. The paratrooper's obsession with eliminating traces (squelching stories, stopping history) is not, to paraphrase Baudrillard, the traditional housekeeper's obsession: that everything be in its place and that everything be clean . . . that was moral, this is functional.[77] At stake is the gradual development by General Massu and his troops of something Massu sought to distinguish as "functional torture"—something comparable to the medical interventions of a surgeon or dentist—as opposed to the premodern, "artisanal" torture practiced in other wars and thus far in Algeria.

Torture, as Roger Trinquier, chief of the "action/information" section, liked to point out, "is something which could be organized."[78]

After November 1954, when French recruitment leapt from 60,000 to 500,000 troops on Algerian soil—the largest deployment of French troops outside of France since 1830—it had to be; torture became mass-produced: "From this period on the practice of torture became so general that it constituted a problem for the whole mass of young Frenchmen called up for service in Algeria. Still more important, the type of war being waged in Algeria could not be carried on at all without the simultaneous use *both of torture and of the mass of young conscripts drawn from the mother country.*"[79] Generalized and routinized, conducted on a mass level after 1954, torture underwent a further noticeable permutation after 1957 when the military took over police duties during the Battle of Algiers: "What had been at the outset an improvization rapidly became a veritable institution with its appropriate structures, its executives [*cadres*], its executors [*executants*], its panoply of accessories and its rules for functioning."[80] The institution was complete with schools [*écoles de formation de cadres*] for the training of experts or "specialists;"[81] instruction in these schools focused on producing a torture that was "clean," which is to say exercised without sadism and without leaving visible traces.[82] Techniques and equipment were standardized: "Torture became in 1957 a daily and almost banal practice. It functioned everywhere. . . . As for techniques, these hardly varied . . . suspending the body . . . and above all the bathtub and electricity."[83] "Everywhere in Algeria, no one denies it, veritable laboratories of torture have been installed with electric bathtubs and everything that's necessary [*tout ce qu'il faut*]."[84]

A further Taylorization of the process ensued with the clearcut constitution of a separate managerial apparatus:

> The organisms in charge of "information" took on a quasi-autonomous structure after 1957 and the Battle of Algiers . . . the birth and development of the Center of Interarmy Coordination, completed by the *Dispositif operationnel de protection.* This system represented a great advantage for the au-

thorities. . . . Torture was practiced in a closed circuit, in carefully chosen places, with a highly "qualified" personnel of an exemplary discretion. Career army personnel knew of its existence, but all of the army didn't torture. It was content to accept this method, allowing the specialists to dirty their hands.[85]

A full-fledged, French-style industrial organization was in place, with its pyramidal hierarchy—but here, once again, in keeping with the distorted mirroring of French institutions by institutions in Algeria, it is the specialists, the *cadres,* who "dirty their hands," who engage in the "hands-on," menial labor, processing the human material, performing the distasteful tasks that had, in an earlier moment in the war, been left to the *harkis.*[86] The transition from artisanal to industrial activity is recalled in an enlisted man's testimony: "I was given responsibility for information at the level of a company, which is minimal. It was all played out at a very low level. . . . It was torture . . . well . . . I don't dare use the word because it's a little shameful . . . but, in the end it was artisanal. The DOPs [*Dispositif operationnel de protection*], now they were the professionals."[87] The DOPs or managers divided the city into subsections, "each of which had its own 'sorting center' [*centre de tri*][88] which included a torture chamber."[89] These centers, established in all the major Algerian cities, helped make torture more efficient, and henceforth anything could be justified according to a criterion of efficiency: "The Orleansville center, which was set up in an old barn, and the Constantine center, which was in Ameziane Farm, turned into conveyor-belt establishments where torture was applied with scientific precision."[90]

From his privileged vantage point as human material being processed through one such center—the major one, El-Biar—Henri Alleg is able, as he puts it, "for a month, to observe how the torture factory worked" (p. 113): that is, its division of labor, its productivity, its employees at work. And "work" is the word used by the paratroopers to describe their activity: "When he didn't go on a sortie, Erulin and his men 'worked' on the suspects who had previously been arrested" (p. 116). Torture in El-Biar is simply "the routine order of the day" [*la routine de la maison*]: "The torture went on until dawn, or very nearly. Through the partition, I could hear shouts and cries, muffled by the gag, and curses and blows. I soon knew that this was in no way exceptional, but just the routine order of the day" (p. 87). Time-saving strategies are devised and practiced: "I just had time to see a naked Moslem being kicked and shoved into the corridor. While S——, C——, and the others were 'looking after' me, the rest of the group were continuing their 'work' using the same plank and the magneto. They had been 'questioning' a suspect in order not to lose any time" (p. 59).

The "transit center" at El-Biar for Alleg is the double of the factory—or, as in the Bosc cartoon, the perverse double of a newly Taylorized French notion of keeping house, where household tasks such as "cleaning up the Casbah" and "floor polishing" are performed largely at night: "But it was at night that the transit center really came to life: preparations, suspects, noise. . . . Then all of a sudden, the first cries of the victims cut through the night. The real work of Erulin, Lorca and the others had begun" (pp. 115–116). The real work was keeping house: for what was at stake was the question of who, precisely, was "at home" in Algeria. "Algeria *is* France": the endlessly reiterated refrain; or, its variant, "France is *at home* in Algeria."[91] But how, asked Frantz Fanon in 1959, "are Algerians supposed to, as General de Gaulle has disingenuously invited them to, 'return to their homes' [*rentrent chez eux*]? What meaning can this expression have for an Algerian today?"[92] Especially since, for an Algerian, to be "at home" is to be at the very center of the conflict.

For Henri Alleg, inhabitant of El-Biar, time moves differently; he recounts the growing sensation of losing track of time, of falling out of the calendar. That a torture victim should have difficulty differentiating one day from another is not surprising. But the bored, rote demeanor of the torturers—whose thoughts are unrecorded in Alleg's text—suggests an experience of numbed repetition so removed from the "vertical" temporality of event as to rejoin Linhart's description of days on the assembly line, or Christiane Rochefort's character Céline in *Les stances à Sophie* as she is reluctantly inducted into the repetitive tasks of "keeping house." And the ideal torture victim, one who emerges fresh and unmarked from one session, ready to undergo another, recalls Barthes's description of something—a freshly shined automobile, to be precise—"given the immobility of material on which time has no effect." Torturer and tortured, the relation unchanging, as in a functionalist equilibrium, endlessly reproducing itself. Torture in the Algerian War strove to "leave no traces"—which is to say, to immobilize time, or to function as an ahistorical structural system. Faced with an insurgent force with history on its side, the whole colonial system is forced, in its final moments, to modernize: to construct a systemic spatial structure coextensive with the whole terrain, a structure so smoothly and cleanly functional that it could stop the forward movement of time.

3

COUPLES

THE GREAT DIVORCE

Who was "at home" in Algeria? The persistent formulation of such a question gives some indication of the way the Algerian War appeared to the French. The French felt the war to be both a foreign affair, something far away and extraneous, and a disruption within the very body of French society; these two perceptions were experienced simultaneously by the metropolitan French. Algeria, after all, *was* France; the Mediterranean divides France, so the saying went, as the Seine divides Paris. Algeria's status was different from that of Indochina or sub-Saharan Africa; whereas France's role in sub-Saharan Africa was articulated in terms of a right of property, in Algeria, from the outset, France affirmed a relation of identity. Africa south of the Sahara may have been decreed French territory, but never was it decided that Africa south of the Sahara *was* France. French national consciousness for 130 years had thus developed according to a simple principle: Algeria is France. The statement was reiterated throughout the early years of the war, most notably by the minister of the interior under Guy Mollet, François Mitterand.[1] (The predominance of such a principle in French public consciousness can be gauged by the decision made by the editors of *Les temps modernes* to entitle their November 1955 editorial "L'Algérie n'est pas la France.")[2]

It was this identitarian myth that provided the basis for the apocalyptic tone of French official rhetoric during the war; because Algeria was France, the revolution represented an attack, an aggression against the French nation itself. It wasn't just that a France stripped of its empire would have to resign itself to becoming what the military sometimes called "a quasi-Spanish mediocrity,"[3] it was more serious than that: "In Algeria *la Patrie* is in danger, the very unity of France is menaced."[4] "It is not only our prestige that is at stake, it is our national independence."[5] A common slogan of the late 1950s, "No more French Algeria, no more France,"[6] played a role in creating the conviction among many French that the destiny of France itself was being played out at that very moment not in France but across the sea.

But another set of popular metaphors and figures of speech suggests that, with the emergence of the nationalist movement in Algeria, the relation between France and Algeria was widely held, by the French at least, to be a kind of marriage: a long and abiding "mixed" marriage, with its history of dirty family secrets that should best remain hidden. A divorce, a bitter separation, would expose those secrets to the light of day. Longtime secretary-general of the French Communist Party, Maurice Thorez, provided what appears to be the earliest and best-known example of the marriage metaphor in a speech delivered in Arles and again in Algiers in the late 1930s. Thorez was intent on arguing that Algeria was a "nation in formation," and as such not yet ready for independence. What Algeria and France should strive for was a modern marriage, a "free union": "Yes, we want a free union between the peoples of France and Algeria. A free union certainly means the right to divorce, but not the obligation to divorce."[7]

The marriage metaphor's representation of a free but indissoluble union was reasserted by Edgar Faure when he succeeded Mendès-France in February 1955: "Algeria composes a unity with the Metropole that nothing can compromise."[8] Such formulations succeeded in reiterating the familial rhetoric characteristic of colonial discourse while underlining once again the singular nature of Algeria for the French. Algeria, a "nation in formation," was not, as in other colonialist figurations, a child to the metropolitan "mother country"; particularly after the war began, Algeria was to be viewed as an adult—or rather, a semi-adult, a wife, someone with whom a man cohabits, someone potentially capable of evicting you from your home: "The eviction of France from North Africa" will lead inevitably to "the decline of France and the rending of her very soul."[9] In such phrases the economic and the emotional are coterminous; the combination of property—household—relations with a string of affective ties ("There is not a single Frenchman who does not have a cousin in Algeria")[10] comes together in the prevalent image of French colonial history in Algeria as the story of a troubled but loving *"vie conjugale,"* a "cohabitation,"—that is, a largely emotional or affective relation in which economic considerations were secondary. Algeria, after all, unlike the other colonies, was "a settler's colony [*colonie de peuplement*] where two million Frenchmen ask for nothing more than to live in peaceful harmony with twenty million Muslims";[11] or, in the words of Michel Debré, prime minister under de Gaulle, "France is at home in Algeria, for Algeria is France's achievement."[12] (Again, the prevalence of the cohabitation metaphor can be measured by Fanon's need to rebut it specifically: "The French in Algeria have not cohabited with the Algerian people. They have more or less dominated.")[13] The Algerian revolution was experienced by the French as "the destruction of the household" [*la ruine du ménage*][14] with all the attendant woes accompanying violent breakups—for example, familial dirty laundry might be aired in public. The repressive system of French censorship of information regarding Algerian affairs was justified on the grounds of national security—and on the grounds that one should keep family secrets in the family [*laver le linge sale en famille*].

THE MAKING OF THE NEW FRENCH MIDDLE CLASS

What Jacques Soustelle called the great "rupture between the Sahara and France," the national divorce whose violence and tensions defined the entire period, was transpiring in the midst of a massive postwar French reaffirmation of the couple as standard-bearer of the state-led modernization effort and as bearer of all affective values as well. In other words, the newsreel image of Monsieur and Madame Dupont at the *Salon des arts ménagers,* to take one example among many, comes to be both affirmed and overdetermined at the precise moment when the larger family unity, the "couple" France/Algeria, is unraveling. Many factors govern the production of a heightened image of the couple during this period. Certainly a new ideology of love and conjugality was necessary if the state natalist policy, enacted immediately after the war in the aftermath of de Gaulle's 1945 call to the French to produce "douze millions de beaux bébés," was to meet with any success. The "family allocations" rulings that play such an important role in Rochefort's *Les petits enfants du siècle* are one indication of the state's active role in fostering what we might call a "reinspired" level of reproduction in the immediate postwar period.[15] Second, the transition by Parisians and other French urban dwellers to mass-style practices of consumption from the mid-1950s onward served to interpellate the couple as optimal buying unit as well; through advertising, husbands were solicited in their status as "experts" or knowledgeable ones to purchase large consumer durables; together, husband and wife went about the task of furnishing their new modern household. Here again Perec's *Les choses* provides the purest laboratory analysis of the phenomenon: childless and possessing no visible parents or extended family, the characters Sylvie and Jérôme are reduced to their pure function of embodying the desires of a new, streamlined, middle-class couple, a couple disencumbered of both the anxieties and the privileges of lineage, inheritance, and transmission characteristic of an older, nineteenth-century bourgeoisie. Gone is the thick web of interfamilial intrigue, the violent struggles over patrimony that filled the pages of Balzac's *Comédie humaine*. Modernization created a new, pared-down, and streamlined French household, rendered more compact in many cases by recent migrations from the provinces to the city: "She dreamt, but her dreams had the precision of a well thought out decision . . . a little modern apartment in a new building, near the gates of Paris . . . not to have a place in Paris would mean being buried in the countryside forever! . . . So that she wouldn't despair, they had to have an apartment completely to themselves, not one belonging to M. Donelle or Mama Donzert [the in-laws], but theirs. A couple who lives with their parents . . . no." (Elsa Triolet, *Roses à crédit,* p. 108). In exchange for the lost interfamilial ties—Martine's family and in-laws in the countryside in *Roses à crédit*; the grandmother and sister-in-law Elise leaves behind when she moves to Paris to work in the Citroen factory in *Elise ou la vraie vie*—the city offers a new compensatory myth of romantic love and emotional fulfillment within the couple. Edgar Morin, writing in 1962, noted this new concentration or reduction of emotional ties to the unit of the couple: "The totality of affective attachments, which were previously diffused in a number of interfamilial relations, . . . now tends to become concentrated in the couple."[16]

But the "couple" was not only important on the literal reproductive level of providing the next generation of French workers, nor on the more immediate level of providing the current generation of French consumers. It was also crucial in helping define a peculiarly French identity in the midst of the "great rupture," and in the middle of a Cold War standoff as well.

"Neither Ford nor Lenin," the French road to modernization—viewed by the postwar reformist avant-garde as necessary in the name of both progressivism and nationalism—had to thread its way between a vision of communist totalitarianism on the one hand and United States economic and cultural imperialism on the other. Popular images attached to the Scylla of cutthroat American opportunistic robber-baron individualism, on the one hand, and the Charybdis of gray, standardized, rote Soviet mass society, on the other, were equally anxiety producing. During the late 1950s Pierre Poujade was fond of inveighing against the two foreign assaults on the unity of France, the American capitalist one defined by international finance and the communist one *"à la sauce tartare"*;[17] French society for the Poujadistes was equally menaced by American industrialization and Soviet collectivization, both of which would standardize away the unique quality of French life: "Tomorrow . . . we will all have the same shirt, the same suit, the same pair of shoes. We will buy all these things from the same automatic distributer. . . . They want to turn us into a bunch of robots."[18] Poujade's was but an extreme expression of a popular fear common to those who sought shelter behind the watchword of "quality"—a characteristic held to define both French products and the French way of life (the latter embodied by those three stalwart individuals: the peasant, the artist, and the individual entrepreneur). "We must remain the nation of quality"[19]; in the antimodernization discourse surrounding French car manufacturing, for instance, the uniquely French knowledge of *la belle mécanique* was held up to counter the introduction of any large-scale mass production techniques. What was true of the products extended to the producers as well: French happiness could not be embodied by the atomized, alienated, suburban "to each his ownist" U.S. commuter portrayed in the American sociology texts translated during those years (a French translation of David Reisman's *The Lonely Crowd* appeared in 1964 with a preface by Edgar Morin). Nor could the numbed and routinized Stakhanovite worker, driven to exceed all possible production expectations, or the equally faceless Soviet bureaucrat, provide great inspiration. Happiness was to be found neither in the solitude of the American model nor in the communitarian Soviet one: French happiness, immortalized in the quantity of boy-meets-girl films of the era and in the public images of various prominent couples from all walks of life—Sartre and Beauvoir, M. and Mme. Poujade (to whose marital life

FIGURE 3.1 Simone de Beauvoir and Jean-Paul Sartre, Dakar 1950. Courtesy Bibliothèque Nationale.

FIGURE 3.2 M. and Mme. Poujade, March 1956. Courtesy Roger-Viollet.

FIGURE 3.3 Jean-Jacques Servan-Schreiber, François Mauriac, and Françoise Giroud at *L'Express* in 1954. Courtesy *L'Express*.

FIGURE 3.4 Yves Montand and Simone Signoret. Courtesy Roger-Viollet.

Paris-Match devoted all four of its January 1956 issues, complete with extensive photo spreads), Giroud and Servan-Schreiber, Montand and Signoret—as well as in the heavily reproduced "couple typique" of the type interviewed by Marker in *Le joli mai,* or represented as M. and Mme. Dupont in the newsreels, or interpellated as readers by *Elle* magazine quizzes such as "Etes-vous un couple idéale?" ("Are you an ideal couple?")—such images implied and produced the couple as bearer of the totality of affective values.

The completely soldered-together subjectivities of Sylvie and Jérôme in *Les choses* provide the fullest representation of the myth. At no point does Perec allow an individual—say, gendered—identity to emerge for Sylvie or Jérôme; yoked together in an invariable "they," the two characters form a single unit whose definition is provided by the fact that they both want to buy the same things. Deprived of any individual subjectivity, Sylvie and Jérôme are also dispossessed of any communitarian identity larger than themselves: without family and with only a "skin-deep" political involvement that arises momentarily in conjunction with the Algerian War (theirs is "a virtually automatic allegiance to moral imperatives of a very broad and unspecific kind" [p. 72]), their "community" is represented above all as an absence, taking the form of the endless series of consumers who fill out their marketing surveys, and their equally interchangeable, moviegoing friends "almost all of [whom] belonged to advertising circles" (p. 44).

Rochefort's *Les stances à Sophie* and Beauvoir's *Les belles images* begin the process of mining the myth of the couple from the interior; both novels end if not with divorce at least with a strongly established and distinct female subjectivity within the couple. Triolet's *Roses à crédit* is another tale of failed synthesis, in this instance between a maniacally modernized wife installed in Paris "with desires made out of plastic and dreams of nylon" (p. 271), and a husband whose inherited, familial métier—rose growing—gives him roots in some older organic or village temporality that renders him finally incapable of leaving the farm. The novel ends in divorce and violent death: "The absence of a bathroom at the farm had decided their life in common" (p. 193). But fiction enacted the fictiveness of the couple in more subtle ways as well.

In the novels I have been discussing by Rochefort, Beauvoir, Perec, and in the 1966 *Les signes et les prodiges* by Françoise Mallet-Joris also, the couple becomes the site of a remarkable new degree of anxiety surrounding issues of interchangeability and standardization. To take Perec first, it is only the narcissism of small differences that separates Jérôme and Sylvie from their cohort of fellow couples in the marketing world of *Les choses:* "Too sophisticated to be perfectly similar to each other, but probably not sophisticated enough to avoid imitating each other more or less consciously, they spent a large part of their lives swapping things" (p. 44). In the chapter devoted to Jérôme and Sylvie's relations with their friends, the "they" of the couple expands effortlessly to comprehend all the others without distinction, all of them "delighted by the likeness of their different histories, the sameness of their points of view" (p. 45), and all indebted for these opinions and points of view to "that ideal couple" (p. 45), *L'Express* and *Madame Express:* "They would dream, half aloud, of chesterfield settees. *L'Express* would dream with them" (pp. 47–48).

Characters in the novels by Beauvoir, Mallet-Joris, and Rochefort, endowed with at least the semblance of an interiority absent in Jérôme and Sylvie, suffer such interchangeability with considerably more anxiety; indeed what inner life they are represented as having seems to serve no other purpose than to render them cognizant of the anxiety that they have become an abstraction. To have access in whatever degree, in other words, to an individual psychology in these novels is to experience that individuality as under relentless assault, to have the enduring sensation of unreality peculiar to feeling devoid of singularity. Thus Beauvoir's narrator Laurence has the habit of gazing at her husband and pondering the question, Why him and not someone else? And when she does so, she imagines the same thought passing through the heads of hundreds of young women: "Why Jean-Charles rather than someone else. . . . (Some other young wife, hundreds of young wives were at that moment wondering why him rather than someone else)" (p. 168). A friend of hers does little to relieve Laurence of her uncertainty by pointing out to her that Laurence's husband and her lover could easily be the same person: "They're both guys with pretty ways and white teeth; they know how to say the right thing and they both plaster themselves with *after-shave*" [in English in the French] (p. 86). Céline's marriage in *Les stances à Sophie* gains her entrance into a purely formal social world made up of identical "couple" households: "We were having dinner with the Bigeons and the Benoits. The next time with the Benoits and the Duplats, or the Duplats and the Bigeons. They always came in pairs. 'Couples' [*jeunes ménages*], they're called" (p. 101). The main character in Mallet-Joris's 1966 *Les signes et les prodiges* is haunted by the image of the mass-produced couple; her nightmare image is that of "ten couples, or twenty, more or less well put together, performing at the same moment, in beautiful rooms on that June evening, the identical gestures."[20] The opening pages of *Les belles images,* published in the same year, feature the same dull anxiety; at a garden party at her mother's country home, Laurence is convinced that the bucolic family scene being played out before her eyes is not only theoretically reproducible but is *in fact* being repeated at that very moment—the identical inflections, the identical postures—in any number of gardens in the immediate vicinity: "At this very moment in another garden, wholly different but exactly the same, someone else is saying these words and the same smile is forming on another face: 'What a wonderful Sunday!' Why do I think that?" (p. 10). The terror of such reproducibility preoccupies Laurence through the course of the novel.

How are we to account for the anxiety registered in these novels: the fear that not only one's accoutrements and surrounding environment but, by extension, one's friends, family, one's very body and facial expressions, have become reproducible? The couple, that way of life that was supposed to provide the vehicle for the French to avoid the standardization associated variously with North American or Soviet reality, has become instead the very mark of standardization. Fictional couples, abstractions made flesh, sentenced to interchangeability, are seen fearing their own fictiveness, trying to avoid coinciding with, and therefore becoming, images, fictions. In *Les signes et les prodiges,* characters discuss an eerie sensation of feeling themselves to be enclosed within the decor of an American film—with no exit!—then wonder if this would not itself make a good topic for a film.[21] It is important to reiterate here that in each of these novels, the major characters—journalists and filmmakers in *Les signes et les prodiges,* market researchers in *Les choses,* advertising executives in *Les belles images,* and *planificateurs* in *Les stances à Sophie*—are in the business of image production, key players in the work of French modernization. In *Les belles images* the degree of the characters' embroilment in a qualitatively new level of reification and repetition becomes a properly *narrative* problem: because Laurence is so thoroughly implicated in that discursive world, Beauvoir allows no formal indication in the text of the origin of discourse. As such, the relation between Laurence's voice and the external narrating voice is blurred, and something resembling Laurence's own point of view or distinct individuality is for the most part submerged beneath (or interchangeable with) the din of all the others: the clichés that echo in her mind are indistinguishable from her own thoughts. Is any given sentence in the book something Laurence is thinking, something she is overhearing, or the memory of a remark her schoolteacher once made to her?

Laurence is most herself, if I may put it that way, when she is hallucinating her own repetition, when she is registering, that is, something akin to class horror. For on one level, the couples that inhabit the novels of Beauvoir, Perec, and Rochefort are interchangeable because they indeed *are* exactly the same: all *jeunes cadres,* all healthy and brave representatives of the new streamlined middle class, the spearhead of

French modernization: the technocrat generation. As such, we might read these novels as nothing more (and nothing less) than the attempt to provide what Perec, referring to *Les choses,* called "a barely heightened description of a particular social set,"[22] or, in other words, the attempt to define a class. Following Derek Sayer, though, we must understand the work of definition or description in this case to be a narrative, historicizing project: "To define a class—or any other social phenomenon—is, in the final analysis, to write its history."[23] The representational dilemmas facing each of the novelists are then several: How can the world of this new middle class—sealed off, secure, privatized, and endlessly self-reproducing—be made to appear at once normative and a partial (class-based) reality? How, in other words, does one represent a kind of newly omnipresent capitalist culture—what Perec calls "the joy of deep-pile fitted carpets"[24]—as omnipresent while at the same time evoking the uneven development, the exclusions, on which such culture is necessarily based? And how does one write the history of a class whose privatized and peculiar mode of social existence is predicated on the very denial of history?

But a more basic problem arises when we look more closely at the social origins, trajectories, and milieus of the characters in these novels. In what sense can it be said that the novelists are writing the history of the same class? Laurence and her family in *Les belles images* are wealthy Parisians, equipped with country homes and luxurious automobiles; Laurence's mother's lover "runs one of the two biggest electronics companies in the world" (p. 23). Céline in *Les stances à Sophie,* on the other hand, is an orphan from a working-class background who finds a job toward the end of the novel in a striptease joint. Jérôme and Sylvie and their cohort in *Les choses* are "the new generation . . . technocrats on the way, but only halfway, to success. Almost all of them were from the lower middle classes, whose values, they felt, were for them no longer adequate" (p. 50). Martine in *Roses à crédit* has managed to transcend her rat-infested, fatherless childhood of dire rural poverty to become an *employée*—one of the thousands of poorly educated young women, mostly from the provinces, who come to Paris to work in low-paying jobs in stores. It is only by stretching the very conception of the middle class that such a varied group of social actors and lifestyles could be said to belong within it.

But this is precisely the history that these novels, taken as a group, narrate: the consolidation of a massive French, urban middle class in the 1950s and 1960s under the auspices of capitalist modernization. The interests of modernization lie in unifying the diverse middle classes, and this involves leveling the difference between an older middle class and those two politically dangerous classes: the (potentially militant) working class and the traditional petite bourgeoisie, breeding ground for the various forms of fascism. If the rural peasantry—whose precipitous decline in numbers during this period constitutes one of the most palpable social phenomena—subsists, as we will see, in a temporality completely alien to that of modernization, the working class and the traditional petite bourgeoisie, on the other hand, live close enough to modernization's rhythms to potentially disrupt it—the latter, by moving too slowly, and the former, by possibly not moving at all. Workers, after all, have been known simply to stop work. And the rugged individuals of the petite bourgeoisie, who found a brief collective embodiment in the figure of the postwar traveling salesman Pierre Poujade, direct their gaze backward: toward the masochistic patriotism of Vichy and the paradise of small, independent producers, away from the corrupt and monstrous Paris with its swarm of experts and bureaucrats. The French petite bourgeoisie revolt in the mid-1950s—"the most surprising story of the epoch"[25]—but not in order to overthrow anything, to move history forward: they revolt in order to live "like before." During its brief but significant heyday, Poujadism dug in its heels and announced itself incapable of crossing the historical threshold of modernity; it sought not to found a world but to restore one. Poujadism, then, is the inversion of bureaucracy; it lies in a distorted mirror relation to the technocrat world of Jérôme and Sylvie.[26] The Poujadist couple—for Poujadism too apotheosized the couple—installed behind their shop counter, constitute a strict domestic and professional association. The daily life of the family and the business are united within the boutique: during Poujadist rallies the wife minds the store. A menace to the couple would affect the commercial equilibrium and prosperity as well.

Claire Etcherelli's *Elise ou la vraie vie*, the best documentation of the divisions within, and aspirations of, the working class in the early 1960s, represents very little of the "embourgeoisement" of the working class that would preoccupy social theorists, such as Mallet and Crozier, writing around the same time. The novel chooses instead to represent the failed union or synthesis between the French working class and the immigrant Algerian population in Paris during the late 1950s in the form of a doomed love affair between a provincial French woman and an Algerian car worker. Yet despite the very deep divisions that prevent French workers from joining with Algerian nationalists for any other reason than self-interest ("French boys are dying down there"), or Algerians from joining with any of the standard worker organizations in place in the factory, both groups (and the uneasy, tiny alliance between them embodied, briefly, by the lovers) manage at various points to shut the factory down.

But even such a divided working class as the one represented by Etcherelli poses a threat to the smooth progress of capitalist modernization. Full-scale modernization of the economy and of society can be carried out only when the working class and the petite bourgeoisie have been eliminated or altered beyond recognition. Consumption, as Michel Aglietta points out, must be organized and stable, while remaining compatible with the supposedly "free" relationships of commodity exchange. This was achieved by generalizing throughout the working class the social structure that was the condition for its cultural integration into the nation, that is, the small family unit and household, and the formation of new expenditure habits based on that norm.[27] How does the realist fiction of the period represent this transformation? In the novels we have been discussing that focus on the consolidation of the "new bourgeoisie," the work of alteration and elimination is accomplished principally by way of the entity Perec refers to as "the ideal couple": *L'Express* and *Madame Express*. Magazine reading, in other words, serves to unite the various characters into a coherent group, within each novel and across novels as well: despite their different levels of education and their varying access to credit, the rhythm of the daily—or rather weekly—lives of Jérôme and Sylvie, Martine, Laurence and her mother, and Céline is constituted by the repetitive, secular ceremony of magazine reading.

Television plays a very small role in these novels, in keeping with the quite limited access (compared to the United States) of French people to television in the 1950s and well into the 1960s. Personal histories of the period show, for example, that the photographs of the Algerian War that appeared in the magazine *Paris-Match* provided far more French with their visual images of the conflict than did either television or cinema newsreels. Martine, growing up in the late 1950s just sixty kilometers from Paris, has never been to the movies or seen a television. (Later, in her high consumer period, she buys a set and even appears on a game show to try to win money with which to pay her debts.) Laurence's family owns a television but never watches it; Perec and Rochefort make no mention of television whatsoever. It seems clear, though, that we are at the edge of visual culture in these novels, and that in French society as it is represented by Triolet and Beauvoir especially, the glossy visuals of its photography serve to prepare the magazine reader for the saturation of visual images to come.

Thus it is magazines that provide the welcome impetus for Martine in *Roses à crédit* to pass "from one universe to another" (p. 47)—that is, from the filthy squalor of her childhood shack to the sparkling and antiseptic home and employ of her adoptive mother Madame Donzer, "with its enamel, linoleum, pale oak, soaps and lotions" (p. 47). The

The Making of the New French Middle Class

chapter recounting Martine's miraculous transition is entitled "On the Glossy Pages of the Future"; in the hairdresser's shop Martine is initiated for the first time into the habit of poring over fashion magazines "where one saw beautiful women, and nylon on every page, gauzy creations for the day and night, and suddenly, taking up a full page, an eyelid with wonderful lashes or a hand with pink nails" (p. 67). A few years later, when she is well established in Paris, employed as a manicurist in the prestigious *Institut de beauté,* she is, in the narrator's words, "living in the interior of the satin pages of a luxurious magazine" (p. 75).

Martine is the least critically astute of the characters and the most given over to the promises and enticements of the glossy pages whose very glossiness becomes a fetishized quality in and of itself. As a child Martine was teased by her family and called a magpie because she was drawn to shiny, sparkling objects; the glossiness of the page is enough to lure a girl who has never been to the movies and never seen a television. Martine uses women's magazines as a script and a blueprint, transforming herself in accordance with their directives regarding appearance and household maintenance—until, that is, she is punished by her author for her naiveté and made to return to her class origins to die a gruesome death in the rat-infested forest. Céline in *Les stances à Sophie* experiences the same magazine directives as a nagging imposition or aggression: "Read *France-Femme* and you'll know how to live," she comments to herself ironically; a few moments later a saleswoman tells her, "Purses and bathing suits are also tango-colored this year. I don't know if you noticed, but if you look in *France-Femme* you'll see that it's the rage on every page" (pp. 65–67). The heroine of *Les signes et les prodiges,* described as having too often visited [trop fréquenté] the world of women's magazines, has developed a strange "modern" symptom: deprived momentarily of the emotional didacticism of the prose in women's magazines, "where one always knows whether one is happy or unhappy: 'He makes me suffer . . .' 'I would be perfectly happy if only my husband's job were . . .'" (p. 304), she can no longer gauge her own emotional climate.

Magazine reading in *Les choses* and *Les belles images* emerges as the chief contributing factor to the "derealization," the sentiment of reproducibility or cloning I discussed earlier. Thus in *Les choses,* for example, it is the chapter that treats Jérôme and Sylvie's relations to their friends that includes a lengthy discussion of *L'Express;* gradually we realize that it is by way of the shared habit of reading specifically that magazine ("they bought it or, at the least, borrowed it from each other," [p. 46]) that the "they" of Jérôme and Sylvie is able to extend out beyond the couple-dyad to encompass a whole "horizontal comradeship"—the phrase is Benedict Anderson's—of like-minded companions. The magazine supplies them with "knowing winks," "specialists who knew what they were talking about," and "those little details which mean everything"—in short, "the middling tastefulness which, for men as for women, is what is so right about *Madame Express* and, by repercussion, about her husband too" (p. 46). But despite everything *L'Express* has to offer, the relation of Jérôme and Sylvie to their magazine of choice is fraught with ambivalence: like Martine, they aspire to the bourgeoisie-without-blinders lifestyle it promotes ("*L'Express,* and that magazine alone, matched their art of living" [p. 46]). But like Céline, they fight its influence. Being wary, disagreeing with its positions, and ridiculing its jargon enables them to go on buying it and reading it: "Their contempt for it kept their consciences clear" (p. 47).

Laurence's world in *Les belles images* is one where low-slung coffee tables at arm's reach are piled high with the latest issues of *L'Express* and *Marie-Claire,* and where country homes—remodeled farmhouses, to be precise—are indistinguishable from the illustrations contained in *Votre jardin.*[28] At a party Laurence becomes dimly aware of the role played by journalism in the nagging feelings of standardization and loss of individuality she has been experiencing when she overhears a man spouting an opinion she has recently read in a magazine: "Why I read that not long ago in a weekly. Since she had taken to looking at magazines again Laurence had observed that people often reproduced the articles in their conversation. Well, why not? They had to get their information somewhere" (p. 116). Even authors have to get their information somewhere. "I wrote *Things,*" says Perec, "with a pile of *Madame Express* beside me, and to wash my mouth out after having read too much *Madame Express,* I would read some Barthes."[29] The language of *Les choses,* according to Perec, was meant to function quite differently from Robbe-Grillet's neutral or surfacy language: "*Fitted carpet,* for instance: for me, that phrase conveys a whole system of values—specifically, the value system imposed by advertising. So much so that you could say that, in places, my book *is* a piece of advertising copy . . ."[30] And Beauvoir: "It is a society that I keep as much as possible at arm's length but nevertheless it is one in which I live—through papers, magazines, advertisements and radio it hems me in on every hand . . . what I wanted to do was to reproduce the sound of it. I ran through the books and magazines in which this sound is recorded."[31]

A magazine is flimsier than a book and more durable, more substantial, than a newspaper. The intermediate object status of the magazine—in terms of its physical properties, price, and prose style in the range of print culture makes it more *shareable* than either the book or the newspaper. Picked up and put down more readily than a novel and by more readers, lying on tables in doctors' offices, abandoned on trains for the next commuter, passed from neighbor to neighbor more frequently than a newspaper, each copy of a magazine tends to be read by more than one person. The growth of the magazine industry after the war bears some relation to the expansion under capitalism of what Lefebvre calls "constrained time": in this case, the frequent periods of waiting or transition—appointments, commuting—that break up a day and which the portable format, easily digestible prose, and manageable article length of a magazine can help fill. And the format of the "weekly" echoes the rhythm of the work week.

What made *L'Express* by far the most influential and widely read of the weeklies, what made it the supreme vehicle for capitalist modernization in France, was its successful intertwining of three narratives. "Three stories or histories [*histoires*] are intermingled," writes Françoise Giroud, "The story of a man and a woman, the story of a magazine, and the story of a group of people who wanted with all their might to make France 'take off.'"[32] The man and the woman are of course that other "ideal couple," Giroud herself and Jean-Jacques Servan-Schreiber, whose perfected Franco-American lifestyle lent metonymic luster to the pages of *L'Express* and *Madame Express*. The group of people who wanted France to take off was the band of well-educated, proto-technocrats like themselves, anxious to leave the crises of the après-guerre period and decolonization behind them, to consecrate their efforts on the economic renovation of their country, and to benefit from the general improvement in the standard of living that followed from it. In the early 1950s that group, more specifically, consisted of the people who wanted to and succeeded in bringing Pierre Mendès-France, "the only politician capable of accomplishing the modernization . . . of French society" to power.[33] Three intertwining histories—that of a couple, a commodity, and a class—coming together to make a new France.

In his widely read book on nationalism, Benedict Anderson argued the role played by print culture in establishing a national community: anonymous readers are connected together through print, establishing a "deep, horizontal comradeship" that becomes the embryo of a "nationally imagined community."[34] Anderson is discussing the relation between founding novelistic texts written in the vernacular and the early stages of creole-nation formation—not our own period, which transpires, roughly speaking, in a dozen years after electricity and before electronics, after the arrival of television but before many French possessed one, in a nation intent on reformation after its colonial "divorce," not formation.[35] But something like Anderson's "deep horizontal comradeship"—a deeper one, in fact, or at least broader given increased literacy and greater access to the magazine-commodity—is at work in the recurrent image of magazine readers in these novels, engrossed or distracted, but performing nevertheless the weekly ceremony replicated simultaneously by thousands (or millions) of the new, enlarged middle class.[36] Through this ceremony, presided over by "the ideal couple," both a national and a class identification is forged around the project of constructing the new France.

NEOBOURGEOIS SPACE

For these authors, establishing and representing their characters' relation to the mass-produced commodity of the magazine provides the first solution to the problems of history writing I raised earlier. The second solution involves each of the novelists recognizing a particular structural necessity at the level of plot. Toward the end of each novel, well after the claustrophobic middle-class Parisian environment is established for the reader, a voyage takes place outside the hexagon, to allow the characters an encounter with the thoroughly and decidedly unmodern: Tunisia in Perec, southern Italy with its eroticized peasantry in Roche-fort, austere Greece with its poverty-stricken villagers in Beauvoir. Yet one need not even leave the hexagon: France's own rural backwardness, a mere sixty kilometers from Paris, constitutes the unmodern in Triolet. The importance of such voyages is less spatial than temporal: their function is to illustrate, briefly, the Blochian theme of nonsynchronicity, the idea that not all people live in the same time with others. Econom-ically, ideologically, and culturally, peasants—whether French or Tuni-sian—inhabit an older time; they remain what Bloch would call a "distorted remnant" of an older economic consciousness in the midst of the flexible, capitalist century, in possession of their own means of production, and tied to a seasonal, cyclical temporality.[37] Peasants, like any other historical residue, enter the perceptive sphere of the already modernized as but one of two things: bothersome or picturesque. In each of the novels the unmodern remains unassimilated, nonintegratable, incompatible with the modernized, technocrat milieu; and this incommensurate distance is represented in part as a dilemma that occurs at the level of consumption. Shopping in Tunisia, "in this foreign city where nothing was theirs" (p. 104), Sylvie and Jérôme are distressed to find that the objects in the bazaar there "don't speak to them," don't beckon, call out to them to be purchased like the objects in Parisian shop windows: "In the end there was nothing to attract them in this sequence of poverty-stricken stalls, of almost identical shops, of cramped bazaars" (p. 106). . . . "They sought signs of complicity all around. Nothing answered their call" (p. 111). . . . "They bought nothing . . . because they did not feel drawn to these things. None of them, however lavish they could occasionally be, gave them a feeling of wealth" (p. 116).[38] In Greece, Laurence buys a series of eggs, each one of which turns out to be rotten.

Such encounters serve to render the original Parisian milieu, in retrospect, even more impervious to that which is unlike itself. In *Les belles images,* for example, the Paris environment is so socially seamless that it is with some surprise that the reader realizes that a minor (but in the end quite significant) character, the little Jewish friend of Laurence's emotionally troubled daughter Catherine, is in fact the only character possessed of any difference whatsoever to inhabit that world, and that much of the novel's plot has revolved around attempts to eliminate this friend and what she represents from Catherine's environment. For after much parental interrogation and even psychiatric intervention, it emerges that Catherine's emotional breakdown has been precipitated by the evidence of that which everything in her upbringing has sought to deny: the existence of lives lived differently from her own—specifically, in the image provided by her friend Brigit, the knowledge of the existence, on the other side of the world, of assembly lines of "young girls who put rounds of carrots on fillets of herring all day long" (p. 98).

In the world of *Les belles images* Jewishness or any other kind of otherness leads then inevitably to the reality of uneven development, to the whole world of production and its deprivations—alienations that remain qualitatively different from the *malaise* and anxiety of Laurence. At stake, then, are two kinds of repetition: the dominant one, the repetition of the image and Laurence's fear that she has become one of her own advertisements, and the other that the first seeks to shut out, and thus must be set necessarily offstage, seen only as the material of a child's nightmare: the repetition of the assembly line. Production as nightmare: the repressed world of labor has, it seems, reached a qualitatively new level of invisibility in these novels. Like the new Citroen at the *Salon de l'automobile* analyzed by Barthes in *Mythologies* whose fetishized perfection is attributed to its erasure of any noticeable joints or riveting-together of its panels—its seamlessness, in short—tangible signs of fabrication or effort constitute the obscene in this object-world. Laurence is continually aggravated by the safety pin that holds together a seam in the little Jewish girl's skirt; she can't rest until she's shown her the right way to fasten the skirt so that the pin won't show. Sylvie and Jérôme, dissatisfied with their tiny apartment, dream that they will return from their vacation to find it transformed into a spacious, comfortable space complete with "an efficient and unobtrusive heating system, invisible electrical wiring"; they prefer meals that appear as if by magic, showing no signs of having been cooked or prepared: "the slow process of elaboration which turns difficult raw materials into dishes, and which implies a whole world of pans, pots, slicers, strainers, and ovens" (p. 55).

As Laurence's class emerges into novelistic representation in the 1950s and 1960s, at the dawning of image culture, it is peopled by individuals terrified of their own fictiveness—terrified, that is, not as Bourdieu would have us believe, of lacking the right amount of "cultural capital" or status, but terrified, instead, of *having* it—and thereby of being an abstraction. Laurence's oft-repeated phrase, "What do the others have that I don't?" refers more to the sense of being effaced or absorbed into her own abstraction than to any particular material object or possession she is lacking. Her problem is not that she might not measure up to the images of the *Express* lifestyle but that she fears that she actually *has*—and, as such, has become fictional, like the couples who appear in her advertisements for wood paneling and tomato sauce, the fear that she has passed through to the other side of the mirror, become the reification of a reification, been subsumed into an image— this is the anxiety gripping the class called upon to embody the new France. But behind that fear lies an equally strong desire: after all, being an abstraction is but a small price to pay for the privilege and security of living outside of history. And living outside of history—the sensation of the "end of history" peculiar to the dominant class's loss of perspectives in advanced industrial society—constitutes a utopia of sorts, however negative.

The construction of the new French couple is not only a class necessity but a *national* necessity as well, linked to the state-led modernization effort. Called upon to lead France into the future, these couples are the class whose very way of life is based on the wish to make the world futureless and at that price buy security. The French urban techno-couple feels itself to be fictive not only because it is historically contingent and historically new but because it is produced, materially and aesthetically, in its depths and its surfaces, elsewhere—and yet it must *be* France. Its material security is predicated on the conditions of generally and chronically uneven development—the only kind that capitalism allows, and the only way that history reveals itself. At the same time the new couple finds itself in a situation in which the appropriate new habits have been unable to form and must be borrowed, for the most part, from American-style magazines and films. The covers of the first issues of *Elle* used American models whose bodies bore no traces of the hardships of war; early French television advertisements for home appliances were filmed in American suburban kitchens.[39] Like France

---

148

itself, whose prosperity did not prevent it from floundering amid a whole set of economic contradictions—those of an exploiter/exploited, dominator/dominated country, exploiting colonial populations and dominated by American capital—the French middle-class couple secures its prosperity at the cost of the rural backwardness of its own countryside (not to mention its colonies) at the same time that its savoir-faire and props, the accoutrements of its daily life, must be delivered wholesale via Hollywood narrative film and U.S. market dominance.

It is only when such a middle class has become consolidated, when it has begun to take shape as a *national* middle class, coterminous, so to speak, with the nation itself, that a new logic of exclusion like the one operating in Perec and Beauvoir can begin to prevail. The new French urban couple of the economic boom years presages a homogenous national class that is no longer, properly speaking, a class. Having eliminated or absorbed the two "dangerous classes," the potentially militant working class and the traditional petite bourgeoisie, seedbed to the various fascisms, having distanced itself from its colonies, modernization leaves in its wake the "broad middle stratum," the consensus at the center. When modernization has run its course, national subjectivity takes the place of class: one is French or not, one is modern or not. Differences in culture or nationality are reinterpreted in terms of various archaisms, that is, as a set of fundamental, naturalized—racial—disabilities. When class conflict or contradiction (both of which imply some degree of negotiability) have been muted or rendered invisible (as in Perec) or situated offstage, the distant drama of childhood nightmares (as in Beauvoir), then nothing remains but the unequivocal "we" and the "not-we": the return to atavistic principles of racial identity and their attendant spatial logic of inclusion and exclusion. The seamless Paris of Perec and Beauvoir is nothing more than the spatial rendering of these new principles of identity.

What novelists like Perec, Beauvoir, and Rochefort—no less than the Situationists performing their urban experiments in Paris during the same years, or Henri Lefebvre progressively recoding his initial concept of "everyday life" into a range of spatial and urban categories—realized, was the emergence of a new image of society *as* a city—and thus the beginning of a whole new thematics of inside and outside, of inclusion in, and exclusion from, a positively valued modernity. Cities possess a center and *banlieues,* and citizens, those on the interior, deciding who among the insiders should be expelled, and whether or not to open their doors to those on the outside.

Capitalist modernization of the 1950s and 1960s created the new image of society as city in part by replacing the dusty, outmoded image of society as factory. Factories were beset with internal conflicts known as class struggle, and dominated by a politics governed by concerns of economic exploitation. But exploitation could not account for the vague, almost nameless anxieties of Laurence or of Jérôme and Sylvie. Their problems fell more under the category of "alienation," itself a spatial term deriving from the Latin *alienare:* to render foreign, to render other, and further, from *alienus* or *alius,* whence comes the French *ailleurs* (elsewhere), alias, and alibi. Dilemmas of alienation highlight the twin poles of location and identity; to be alienated: to be displaced from oneself, to be foreign to oneself.

Once again, the question is housing. Where is the new middle class to live, especially in view of the acute housing shortage and the dilapidation of postwar Paris? To construct the new seamless Paris for Laurence and for Jérôme and Sylvie, more than two-thirds of the population of Paris had to be expelled from the center of the city either by force or by means of successive rent increases.[40] It is in the new social geography of the city, in other words, that we can begin to see the political effects of the yoking together of modernization and hygiene we examined earlier. Over the second half of the twentieth century these effects would become increasingly racial in nature in the form of a kind of "purification" of the social (urban) body (a purification that would find an almost comical reflection in Malraux's decision, under de Gaulle, to "whiten" the city by sandblasting the surfaces of the most famous Parisian facades).

In the early 1950s Paris still showed remarkable unevenness, the wealthy districts of the west coinciding with the *taudis* of the center and the peripheral arrondissements. But between the years of 1954 and 1974 Paris underwent the demolition and reconstruction of a full 24 percent of its buildable surface.[41] Modernity and hygiene served as a pretext for the demolition of entire *quartiers:* Montparnasse, Italie, Belleville, Bercy. The Hausmannian projects of the mid-nineteenth century were the occasion for the first great emptying out of the city's poor. Under the second wave of expulsions, between 1954 and 1974, Paris proper lost 19 percent of its population—about 550,000 people, or approximately the population of Lyon. But that statistic masks what was in fact a profound reworking of the social boundaries of the city as a result of the renovation projects: in those twenty years the number of workers living in Paris declined by 44 percent. They were dispersed to the outlying suburbs, while the number of *cadres supérieurs* increased by 51 percent.[42] As in the nineteenth century the reasons justifying the reappropriation of space were the same: hygiene and security. And as in the nineteenth century, when recently arrived provincial day laborers—the future Communards of 1871—labored on the urban renewal projects (thus constituting both the instruments and the main victims of the transformation), the twentieth century modernization of Paris employed a large percentage of recently arrived foreign immigrants toward the reconquest of the central areas by the middle class.[43] French immigration policies took their cue from the fluctuating labor needs: a relaxed, open-door policy greeted the onset of the period of economic growth and mushrooming urban renovation in 1954, whereas a new set of stringent immigration restrictions announced the arrival of economic recession twenty years later. French modernization, and the new capital city that crowned it, was built largely on the backs of Africans—Africans who found themselves progressively cordonned off in new forms of urban segregation as a result of the process.

The methods used to create neobourgeois Paris could be predicted on the basis of the renovation projects of the nineteenth century: concentrate within the city limits the centers of public and private decision-making, the research centers and government "think tanks," and offer incentives to employers to move what industrial jobs remained to the outer suburbs. What was less apparent at the time was that such a massive transplanting of industry—with its unskilled, largely immigrant work force—to the outer suburbs would lead to the consolidation of areas of strong minority group concentration on the outskirts of the city. Algerian independence played a significant role in the "*grands ensembles*" boom in the early 1960s—the boom responded to the demands of a mass influx of *pieds noirs* into France and a simultaneous surge in immigration from North Africa. By 1969 one in six inhabitants of the greater Paris region lived in a *grand ensemble*.[44] Paris *intramuros,* peopled by the mostly white upper and middle classes, became in those years what we now know it to be: a power site at the center of an archipelago of *banlieues* inhabited mostly by working class people, a large percentage of them immigrants.

The suburbanization of the immigrant population was not total—significant North African areas such as the Goutte d'or remained vital for the time being, though it would soon be massively redeveloped as a political reaction to the general French perception of the area as a kind of ghetto: "The dread rectangle of which the Goutte d'or is the heart" (p. 229) is how Etcherelli describes outsiders' perception of the eighteenth *arrondissement* in the 1950s in *Elise ou la vraie vie.* And a relatively more welcome influx of Chinese and Vietnamese into the city followed the end of the Indochinese War; these tended to settle in the newly constructed towers and high rises of the modernized Place d'Italie (in part, perhaps, because few Parisians wanted to live there, resenting what could be described as the invasion of *banlieue* architecture into the city proper). Nor was such suburbanization entirely new: the clustering of migrant workers and their families around suburban industrial sites in *bidonvilles* predates their arrival in public-sector high rises. But surveys confirm that the average French person then and now conjures up the image of an Algerian when he or she hears the word "immigrant"— paradoxically, in that the Algerian was, in the time of the empire, a French citizen, as are today (at least for now) his or her children.[45] It is true that Algerian immigration to France rose after the war, owing in part to the hardships of life under colonization but mostly to the introduction of modernized forms of mechanization to Algerian agriculture in 1945. In the 1950s Algerians had priority of entry into France because of the tacit French supposition that they would leave, and because immigration, in general, was needed to provide a mobile and rapidly available work force for the new industrial priorities. In fact, most did not leave: the newly independent Algerian state needed *cadres,* not workers, after 1962. The growing population of Algerians living in France (211,675 in 1954, 350,484 in 1962, 473,812 in 1968)[46] settled in the *bidonvilles* ("shantytowns") on the outskirts (where poor living conditions were compensated for by a degree of social and cultural freedom and by less policing than in the center city) or in the more vigorously controlled Goutte d'or and other of what were called the *îlots insalubres* ("unhealthy blocks") inside the city. It was these areas, the *bidonvilles* and the *îlots insalubres,* that were specifically targeted for aggressive renovation in the early 1960s, at a moment when de Gaulle is reported to have commented to his prefect of the Seine, Paul Delouvrier, that the Parisian region was a brothel *("bordel"),* and that Delouvrier should proceed in "making me some order in there."[47] The demolition of the *îlots insalubres* tended to drive the remaining Maghrebin and black African people from the city center.

Renovation, as Manuel Castells defines it, and as Christiane Rochefort and others lived it, is always aggressive, requiring active state intervention into the urban structure with a view toward changing the function and social contents of an already existing space. The history of the mid-twentieth-century renovations shows the new city to be the logical outcome of capitalist modernization's adroit manipulation of the discourse of hygiene we examined earlier. The areas within the city targeted for demolition, the so-called *îlots insalubres,* were densely populated city blocks that had received their name in the 1930s by virtue of their high tuberculosis mortality rate. In 1954 the *îlots* were still largely intact; more than 100,000 people lived in them, but tuberculosis was no longer a problem. This did not, however, prevent the use of hygiene or sanitation by developers and promoters as the major justification for launching urban renewal operations. References to an improvement in sanitation that would benefit all Parisians served to legitimate most of the renovations undertaken after 1954, most notably the decision taken in 1963 during the de Gaulle regime to remove Les Halles from the center of Paris—the act Chevalier calls "the surgical removal of the heart of the city."[48] That heart, according to the developers involved, was diseased; in their campaign to rid the city of the alleged traffic congestion caused by Les Halles, developers referred repeatedly to the supposed unhealthiness of the markets, invoking "armies of rats" and other invisible filth lurking beneath the structure's surface.

The return of de Gaulle in 1958 and the "reign of the technocrats" that his return announced brought with it a change in the personnel and the philosophy of the ongoing renovation projects. Until then construction was something done by an individual or a company acting by itself. Gaullism introduced a new degree of confusion between the public and the private realms, a confusion personified by the arrival of a new character, the "promoter," who was not a simple builder but instead operated as a kind of intermediary between the banks and the public authorities. "Thus was formed in these blessed years something new to Parisian society or, as Balzac would say, a new social type; this was a world apart, the bizarre herd of those in real estate, the real-estate fauna."[49] A hybrid formation, the promoter or developer was part speculator, part builder, part financier, and a large part publicity relations man; one of his major "promotions" was the myth of construction at any (social) cost. The promoter obtained both the capital and the necessary state authorizations for demolitions and construction, facilitating the interpenetration of state and private interests in what can now be seen as the most massively corrupt era of financial speculation and destruction of the old *quartiers* since the great real-estate speculations and land grab of the Second Empire, depicted so vividly in Kracauer's study of Offenbach.[50]

Inhabitants of the areas targeted for renovation were required or pressured to leave by promoters, often using fraudulent promises of relodging or indemnity. When relodging occurred, it was usually in an apartment in one of the rapidly constructed *grands ensembles*—"vertical *bidonvilles*," as they were sometimes called—built, like Grigny for example, about 20 kilometers outside of Paris, far from any transport lines but next to the autoroute. In compensation for the disruption such expulsion caused in the way of life of frequently elderly and/or immigrant individuals and families, promoters offered the lure of modernization: a bathroom and a modern kitchen. Ironically, in the case of Grigny, such a promise was short lived: slipshod construction and poor building materials caused 450 of the lodgings there to be declared "*insalubre*" in their turn a mere five years after people were relocated there.

Studies of the renovation of specific areas—of Belleville or the rue Nationale area in Paris, or of the Roubaix area on the outskirts of Lille—show that in many cases the very presence of immigrants was used by promoters and the interests favoring redevelopment as an indicator of the need for serious intervention. Despite local differences, the studies all show some degree of convergence between the discourses of hygiene and sanitation, on the one hand, and expulsion of foreigners, on the other. In Roubaix, for example, a decline in the local textile industry (and the resulting need to get rid of its no-longer-necessary foreign workers) coincided with the emerging consensus in the city government around the *insalubrité* of the workers' quarter, its drastic need for renovation.[51] The consensus reached on the dilapidation of the district became a consensus on the necessary departure of the foreigners from the area. Cultural stereotypes took on new vigor, and almost without anyone noticing, the cause of all the evils changed from the dilapidated housing to the presence of the immigrants: limiting their number becomes an act of "social hygiene." It is now the immigrants that by their very being are preventing the redevelopment of the city; it is obvious that they must leave. A total convergence has been reached: the cause of the presence of the immigrants is the unsanitary housing, the cause of the unsanitary housing is its occupation by immigrants.

In the case of a Parisian *quartier populaire* such as Belleville, vast opinion campaigns emphasizing the negative aspects of the targeted zone were launched by promoters before renovation began. The negative aspects were two: lack of hygiene and lack of security, both of which were effectively embodied by the generalized image, in such campaigns, of the "immigrant worker."[52] Since the goal was to liberate the land at the cheapest possible price, the campaigns were followed by attempts to convince the better-off people in the area to better their situation even more by leaving. A shopkeeper or two who closes down can lead to panic in the local population; official documents announcing closures and impending demolitions are frequently terrifying to immigrants who prefer to leave (the area or the country) rather than confront "troubles."

When this procedure is completed and it's time to sell—to Jérôme and Sylvie, for example, or to any of the other *jeune cadre* couples who can afford housing only in the renovated areas of the city and not in the "old money" districts, the *quartier*'s connotations are effectively reversed. Any overcrowding is testimony to the area's liveliness and animation, the small shopkeeper who stuck it out against all odds through the renovation process becomes essential decor, and the worrisome immigrant puts his knife away to become an exotic and picturesque neighbor.

NEW MEN

NEW MEN AND THE DEATH OF MAN

"Societies of a new type are forming before our very eyes."[1] The roughly ten-year period on either side of the end of the Algerian War in 1962 was perhaps the last time that French people were greeted by a resounding chorus proclaiming the "new": Françoise Giroud's coining of the term "New Wave" in 1958 to describe the emergent youth culture in the cities has become in subsequent years the commonplace marker for a historical moment when most people found themselves to be living two or more lives at once, and thus felt the invigoration and fatigue of confronting situations without having the appropriate habits or behaviors firmly in place. Cultural histories of the period have frequently located the energy of the new in its standard avant-gardist location: in the daring break on the part of the group of young directors that became known as the French New Wave with the theatrical and academic cinematic tradition, or in the austere stylistic experimentation of the anti-*école* of novelists grouped somewhat arbitrarily under the rubric of the New Novelists. But an examination of the discursive production of the time shows that the noun most often modified by the adjective "new" was in fact that of "man": in the decade in question, the arrival of the "new man," of a new construction of (male) subjectivity was proclaimed from all sides, celebrated, analyzed, and debated.

But there were, in fact, several of these "new men," depending on where one looked. Their identities and activities at times overlapped with each other and at times remained rigorously distinct, having been defined, in part, in opposition to one another. I will examine here the three principal versions of postwar masculinity, each of which announced itself as "new," and each of which occupied a distinct discursive and spatial site: the *maquis* (those inaccessible areas where armed resistance was organized), the university, and the corporation.

The most overwhelming sense of the "new" emanated of course from the ongoing mass nationalist movements and wars of decolonization. In the year 1960 alone, fifteen African states became independent, in part in the wake of the continuing Algerian struggle against the French. A utopian rhetoric of creation—clean slates and tabula rasa—dominated the writings of anticolonialist writers intent on analyzing the colonial situation and calling for its overthrow. In this sense the opening of Aimé Césaire's *Discours sur le colonialisme* is paradigmatic; in a few brief, elegant sentences the relation between Europe and its colonies is cast by Césaire, in a clear denunciation of the age, illness, and moral corruption of Europe ("decadent," "stricken," "dying") against whose background the colonial peoples can only emerge as historically new.[2] But it is Fanon who goes the farthest in developing the argument that men change at the same time that they change the world. In dialectical opposition to what he calls in *Les damnés de la terre* "the stasis of Europe"[3] comes "decolonization [which] brings a natural rhythm into existence, introduced by new men, and with it a new language and a new humanity. Decolonization is the veritable creation of new men" (p. 36).

Fanon's "new man" is at once a historical, a physical, and a psychosexual phenomenon. In a striking phrase that we will return to later, Fanon equates "the stasis of Europe" with "a motionless movement where gradually dialectic is changing into the logic of equilibrium" (p. 314); Europe, in other words, has succumbed to the static reproduction of functionalist models, to the *posthistoire* of the overdeveloped world. Properly historical or dialectical movement is on the side of the Third World. But that movement is indistinguishable, for Fanon, from the physical, muscular movement of men in action. The historical or dialectical movement that was decolonization ("It is a question of the Third World starting a new history of Man" [p. 315]) is inseparable from the release of muscular tensions accumulated under decades of racist colonial strictures, and the realization of dreams of "muscular prowess" (p. 52). The "natural rhythm" unleashed by decolonization is both a lived, corporeal rhythm and the rhythm of historical necessity.

But the "new man" is also new psycho-sexually, and it is in this sense that Fanon can be read as a theorist of masculinity. His analysis of the colonial situation depends heavily on a Freudian model of castration whereby the (male) castrated colonized subject attains full manhood or "wholeness" through revolutionary solidarity and the violent overcoming and expulsion of the colonizer. This is one of the meanings underlying Fanon's call for "a new direction: Let us try to create the whole man" (p. 313). But despite the fact that the colonizer occupies the place the colonized himself wishes to occupy, the European, in Fanon's view, is not "whole" either; European man is himself maimed: "When I search for Man in the technique and the style of Europe, I see only a succession of negations of man, and an avalanche of murders" (p. 312). The coming to new man*hood* of the colonized involves an unprecedented reconsideration of the entire conceptual question of man*kind:* "The human condition, plans for mankind and collaboration between men in these tasks which increase the sum total of humanity are new problems, which demand true inventions" (pp. 312–313). To a certain extent, Fanon's rhetoric rejoins that of a long tradition of libertarian, existential Marxism like that of Lukács for whom communism was the realization of "authentic humanity" and man's "total personality." Henri Lefebvre too held Marx's concept of the "total man" to be the strongest critical weapon against the reduced "economic man" created out of the reified, instrumental relations of Western, bourgeois, market society.[4] But the vision of Fanon's "new man" was in other ways specifically produced by the colonial situation; to be whole does not mean imitating a European model, even that provided by European Marxism: "We do not want to catch up with anyone" (p. 314).

Fanon's acute awareness of the pitfalls of adopting a mimetic relation to the European model (and particularly to the decadent European bourgeoisie) does not prevent his embrace—for the purpose of overturning and transforming it—of the French bourgeois concept and value par excellence: *l'homme* as the foundation of western humanism. But in France that very concept was coming under severe scrutiny and attack—in part, as a critique of ethnocentrism—at the hands of the second of our "new men": the one that Barthes located in French universities and christened in an influential 1963 essay "structural man."[5]

My purpose in the discussion of structuralism that follows in this chapter is not to write an exhaustive analysis of an intellectual movement that had profound and diverse effects on each of the disciplines in the human sciences—I have little to offer, for example, on the subject of structuralist literary criticism. During its heyday, countless books were written and continue to be written on particular figures within structuralism, and on the upheavals within individual disciplines that marked its passing. Now that the debates internal to the structuralist line of thought have subsided, it becomes possible to see what that line of thought looks like when it is situated in conjunction with both decolonization, on the one hand, and with the state-led modernization effort, on the other. If we cannot argue that the whole "end of history discourse" deriving from structuralism is itself an effect of the material changes in postwar France, we can at least determine the ways in which the two came together in this period, drew on each other and reinforced each other to create an ambiance of timelessness, a depletion of the historical during a time of profound reorganization of lived experience at the daily level.

Structural man, for Barthes, was defined by neither his ideas nor his discourse but by his imaginary, the way in which he mentally lives structure. Compared to the Fanonian "new man," structural man was a disembodied creature, a set of mental processes. Structuralism is an activity only in the sense that it involves a regulated series of mental operations; its goal, an entirely imitative or mimetic one, is to reconstitute an object in such a way as to manifest the rules of functioning of the object. Structural man takes the real, decomposes it, then recomposes it in view of creating the general intelligibility underlying the object; he creates the object's simulacrum. The activity of arriving at this level of general meaning is more essential to structural man than the weight or validity of any particular meaning itself: "*Homo significans:* such would be the new man of structural inquiry" (p. 218). Subjectivity, consciousness, and agency—what passed for *l'homme,* in short, under the now obsolete terms of bourgeois humanism—are effaced to the profit of rules, codes, and structures.

Structural man—embodied, for most observers, by the figure of Claude Lévi-Strauss—was a French invention. But Lévi-Strauss's position at the forefront of the fields of ethnography and anthropology brought structural man in contact with the non-French and the non-West. In other words, structuralist discourse, which was not long content to stay anchored in one discipline but instead became the basis for new possibilities of interdisciplinary "communication" ("a fluid that saturated everything," as one critic put it),[6] nevertheless began as a science of non-Western societies. The product of a strong Durkheimian lineage, Lévi-Strauss combined this genealogy with Jakobsonian linguistics, as well as with his own personal predilection for the science of geology, and headed off for Brazil. His dominance over the discipline of anthropology would extend well into the 1970s.

Still, François Dosse is correct in pointing out that only rarely was the structuralist grid applied to Africa. Lévi-Strauss and the numerous structuralist anthropologists following in his wake showed a marked preference for analyzing the meager remnants of Indian communities in the Americas, victims of genocide and, to a certain extent, societies "frozen in time"—"speculating on dead societies while still remaining safely under the protection of scientific standards"[7] was one contemporary description of Lévi-Strauss's methods—and an equally marked aversion for or silence about the colonial peoples of Africa or Indochina, at that moment engaged in full historic mutation. Structural paradigms, it seems, could offer little understanding of militant peoples steering their own history-in-the-making. Nor, as Dosse suggests, does the previous colonial history's complex imbrication of local customs and European institutions present the neat and hygienic binaries that structural paradigms are built upon.[8] Structuralism's avoidance of the question of decolonization took the form of political lethargy on the part of its proponents as well: one looks in vain for the names of prominent or soon-to-be prominent structuralists such as Barthes, Lacan, Levi-Strauss, or others, as signatories of what was the most important antiwar petition by French intellectuals to emerge during the Algerian War, the 1960 *Manifeste des "121."*[9]

This standoff, or structuralist aversion, for societies in full revolutionary and historical flux can be explained, at least in part, by the various existential distances—racial, national, and class—separating structural man from the "new man" emerging simultaneously in the writings of Fanon. But it is also a conflict over the status of the concept of "man" itself. "We must . . . try to set afoot a new man," writes Fanon in 1961 (p. 316); Lévi-Strauss a year later: "I believe the ultimate goal of the human sciences is not to constitute but to dissolve man."[10] And four years later, in Foucault's *Les mots et les choses,* the future of man as his own dissolution or erasure reached its most passionate expression: "Man would be erased, like a face drawn in sand at the edge of the sea."[11] Precisely at the moment that colonized peoples demand and appropriate to themselves the status of men, in other words, French intellectuals announce "the death of man."

The battle around the word "man"—its embrace by revolutionaries in the colonies and workers at home, its effacement by French structuralists—reached a new level of contradiction within structuralist Marxism, notably Althusserianism.[12] For the antihumanist Marxists who came to the forefront in the mid-1960s, "man" is of course bad because it is nothing but an image that masks the conditions of bourgeois domination. Even the utopian new "total" man favored by existential Marxism was unacceptable to the positivism of a structuralist or "modernized" Marxism like that of Althusser. By the mid-1960s the proclamation of the death of man and the liquidation of the subject had begun to be heard throughout the academy; no serious French intellectual could invoke "man" without blushing, or at least without using scare quotes. And yet the antihumanist or structuralist Marxist found himself surrounded by workers and colonized peoples who did nothing but claim for themselves the status of man, whose slogans, like those of Fanon, were founded on the call for a more *human* society: "we are men, not dogs," "the economy in the service of man, not man at the service of the economy," "socialism with a human face," and so on.[13] People in struggle, oblivious to the embarassment of structuralists, seemed to fight all the harder to be recognized as men.

In the writings of Fanon, Césaire, and Albert Memmi the revindication of the status of man on the part of the colonized is first and foremost a refusal: the refusal to be *qualified* by the masters. To be colonized, for Memmi, is to be constantly qualified or subjected to a series of often contradictory negations and reductions. The colonized is not this, not that: "But who is he? Surely not man in general, the holder of universal values, common to all men. In fact, he has been excluded from that universality, both in word and in fact."[14]

The power that masters give themselves to qualify is a direct expression of property rights over the colonized. To refuse that power—to claim, as in Fanon, the status of "whole man"—is to stop the tendency toward slavery, toward being turned into merchandise, toward being turned into the nonhuman. And the nonhuman, in the colonial situation, is most frequently an animal: Memmi, Fanon, and Césaire have all written at length about the racist "animalization" of the colonized in an elaborate battle of the species whereby "only one of us can be a man."[15] To animalize the colonized is the most efficient way for the imperialist to situate them as the second term, lacking the fullness of the whole, deficient in human qualities. Animalizing goes hand in hand with what Memmi calls "the mark of the plural": each individual colonized is "less than" the bourgeois, whole European man in part because there are "too many" of the colonized, thus the imperialist rhetoric of swarms, masses, hordes, overbreeding, and spawning. Césaire's pamphlet outlines the slow and inevitable process of what he calls the "boomerang effect of colonization": "the colonizer who, in order to ease his conscience gets into the habit of seeing the other man as *an animal,* accustoms himself to treating him like an animal, and tends objectively to transform *himself* into an animal" (p. 20, italics in the original). In the eyes of the colonized the dynamic of animalization turns back on itself; the imperialist, however, continues to view the colonized as qualified, lacking. By qualifying, animalizing, and pluralizing the colonized, the European continues to supply himself with wholeness, humanity, and integrity.

Structuralist Marxists would argue (and did) that by suppressing "men" they were not suppressing "real men." They were suppressing the mask, the image that concealed bourgeois domination; they were concerned merely with "man" as an ideological myth, a philosophical concept with a particular theoretical function. In this way structuralism avoids ambiguity. But what of those historical situations where "man" has actual political effects, where it constitutes a political demand? Structuralist Marxism must ignore those situations (as it ignored Africa) where "man" becomes a rallying cry; it avoids ambiguity at the price of being nothing more than a theory of representation. The distance separating the new structural man from Fanon's new man is the distance separating representational theory from political praxis or agency, or, in Fanon's terms, "the logic of equilibrium" from "dialectic." The structuralist "death of man" was the death of the "new man" as hope for a transformed or utopian future as well: science could have nothing to do with revolutionary action.

CADRES

Fanon's revolutionary *cadre* was the aspiration to a future; he was both present—an activity, a subjectivity, a force—and the wish for a different world. His being, in Fanonian rhetoric, is intimately tied to the being and future of the nationalist movement: "The Algerian nation . . . is at the very center of the new Algerian man. *There is a new kind of Algerian man,* a new dimension to his existence."[16] He both is and will be: Fanon ends *Les damnés de la terre* with this final exhortation: "We must turn over a new leaf, we must work out new concepts, and try to set afoot a new man" (p. 316). In a 1965 text widely read in France after it was translated in the mid-1960s, "Socialism and Man in Cuba," Che Guevara uses an identical rhetoric: "In this period of constructing socialism, we can be present at the birth of the new man. His image is not yet completely fixed, it could never be given that the process is parallel to the development of new economic structures. . . . It is the man of the 21st century that we are creating . . . we will forge ourselves in daily action while creating the new man."[17]

It is this opening onto a future temporality that Fanon's revolutionary *cadre* shares, at least discursively, with his distorted mirror "double" across the sea, the third of our "new men," the *jeune cadre*. We have seen the *jeune cadre's* efficient, energetic style already in the shape and discourse of Jean-Jacques Servan-Schreiber, or in an array of fictional husbands: Laurence's husband in *Les belles images,* Céline's in *Les stances à Sophie,* all well shaven and armed with a steadfast faith in a techno-logically perfected future. But these examples shouldn't lead us to believe that the French *jeune cadre* was a purely metropolitan phenomenon. Since his "birth" coincided with the final days of the empire he appears briefly in the colonial setting as well, the harbinger and facilitator of the tran-sition to neocolonialism:

> If the colonizer appears to everyone to be a washed-up char-acter, not susceptible in the new circumstances of being the official bearer of French and western civilization, on the other hand a new man makes his appearance, discreet but insistant: the cooperative technician. The pretext and justification for his presence is his knowledge and efficiency. He doesn't "get involved in politics" and shows proof of good will and de-votion. Thanks to him the colonial enterprise regains its lost good conscience.[18]

Whether in the colonies or in France, the new, forward-looking corporate *jeune cadre* was the transitory expression of a French society in transition. The product of a series of social mutations, he was de-monized in Poujadist rhetoric as a manipulator of "anonymous and vagabond capital"; as one of those "technocrats, more or less stateless, who broadcast the claim to regulate our destiny."[19] And he was heralded in countless magazine articles and books in the early 1960s as "the prefiguration of the new world," as "the man of the future."[20]

In one such book, the 1961 *Un homme nouveau: L'ingénieur économiste,* Francis-Louis Closon claims to detect on the horizon "a man already in place and moving along rapidly yet appearing to be without tradition."[21] In fact, the idea that either of the two "new men," the revolutionary *cadre* or the *jeune cadre,* had arrived on the scene fully formed and without historical lineage is silly. The Algerian FLN fighter had a complex national and international geneology, extending back through a rich history of Algerian insurgency to nineteenth-century figures like Abd el Kader and El Moqrani, and laterally to fraternal relations with Che Guevara and Fidel Castro's "new socialist man," and to Mao's peasant armies of 1949, not to mention the more obvious pan-Arabic ties, or the very crucial kinship with Vietnamese nationalists. (The Algerian revolution broke out a mere five months after Dien Bien Phu; many North Africans had died in Vietnam, fighting on the side of the French.) As for the *jeune cadre,* Luc Boltanski has traced his national and international ancestry, from his beginnings in the 1930s (the word *cadre* entered French industrial vocabulary around 1936 and was used to describe men whose place in a company couldn't be precisely defined), his consolidation in the corporatist values of Vichy, through to his full-fledged flowering in the late 1950s.[22] That flowering was in fact the successful grafting or composite formation of two previously separate individuals: the engineer and the bourgeois, and it occurred under the careful midwifery of the American productivity missions conducted after the war.

The grafting of bourgeois and engineer is given full narrative form in the character of Daniel, Martine's husband in *Roses à crédit.* Born in the same village as Martine but to a much wealthier family, Daniel will successfully reconcile all of the various nonsynchronicities that bring his wife to a tragic end. It is not surprising that the reader first meets Daniel, the son of a provincial family of rose growers, in Paris, studying to acquire an advanced degree: the *cadre*'s entry into popular discourse coincides with the standardization of higher education in France, the moment when the fact of pursuing an advanced degree ceases to be reserved for a tiny minority. He is studying, however, to be a "horticultural engineer" ("But those two words don't go together!" exclaims one character [p. 106]). In Daniel will be joined a deep historical bourgeois lineage—his family traces its geneology back to the Middle Ages—and the most elite, high-technological, scientific training. Triolet allows

Daniel to negotiate an extremely treacherous historical divide: the one separating the nineteenth-century notion of *métier* (like his father and his grandfather before him he will be a *rosier,* he will inherit the farmhouse, the rosefields, and the set of practices associated with the *métier*) and the more contemporary, American-inspired, or *Mendèsiste* notion of the *career:* he will be a scientifically trained rose grower, a horticultural engineer. He will inherit the *métier* and transform it into a career in a single generation. The newfound ascendancy of the institution of the "career" as the privileged form work takes in overdeveloped and bureaucratic industrial societies is the mark of a change not only in the ideology of work but in bourgeois patterns of accumulation as well. Whereas the bourgeoisie of the past, even the close past of Daniel's father's generation, accumulated savings and land ("the patrimony" to be passed on to future generations), now bourgeois accumulation takes the form of experience at work: the cumulative perfecting of skills, the ascent toward accomplishing more and more highly appreciated and rewarded tasks. The *jeune cadre,* by virtue of his "career experience," is worth more at the end of his professional life, or even at the end of a single day of work, than those workers who "run in place" at work, performing rote, repetitive tasks, the kind of worker who finds himself at the end of his professional life with the same experience he had starting out. But the career is also distinct from the *métier.* Although both presuppose an accumulation of technical experience and both offer some level of security, the job security of the *métier* depends on market demand—this town can only support two bakers—whereas the guarantee of a career position is due rather to the recognition of social *status.* Career positions, in other words, are socially and culturally linked to other situations of privilege and property, and the status acquired by a career has a new and irreversible force in the history of capitalism: the force that comes to be personified in the print and visual culture of the era by the image of the *jeune cadre.* Daniel will have the new forms of power and desire associated with the "career," but he will hold onto an older set of values: nowhere is this more apparent than in the elusive object of his professional desires, the invention of a new rose: "Grafts, hybrids, artificial insemination, the creation of new roses. . . . Daniel was attempting to obtain through grafts a rose that would have the perfume of ancient roses and the form and coloring of modern roses" (p. 98). The perfect hybrid, the perfect synthesis, like Daniel himself, between nature and technique, between city knowledge and country wisdom, between past and present: the sensual pleasure and value of an older temporality associated with the nostalgic possibilities of smell, instilled into the visual design of the present—the visual itself *as* present. Unlike his wife Martine, Daniel will successfully modernize: in one generation he will transform his patrimony into that which cannot be inherited: dynamism, technical competence, and a drive to achieve, the new defining characteristics of the *jeune cadre*.

"The engineer," as Closon puts it, "is not born of a fantasy, but of the force of an efficient world that shows no mercy for those who lag behind."[23] The engineer is both the agent and the product of capitalist modernization just as Fanon's "new man" is both the agent and the product of decolonization ("Decolonization is the veritable creation of new men" [p. 36]). That the same word, *cadre,* could be used to describe the prefiguration of such very different futures shows that the two men share something beyond embodying the force of historical necessity. They are both, as the word itself suggests, completely imbricated in organizations. Both the revolutionary *cadre* and the French *jeune cadre* are "organization men": no one who has seen *The Battle of Algiers* can forget the famous blackboard scene wherein the Colonel Massu, having recently taken over the police/interrogation functions in the city of Algiers, patiently explains the strict pyramidal hierarchy of the FLN organization to his men and to the viewer. The complexly circular narrative style of Assia Djébar's 1962 *Les enfants du nouveau monde,* a novel that sets out to evoke what one character calls "the palpitation of the new world to which she would belong," is another, very different, testimony to the networks of solidarity and organization at work during the revolution. The circularity of the narrative, the enfolding of characters' experiences and viewpoints into a vast, revolutionary mesh, reflects what another character, awaiting her interrogation and emprisonment by French authorities, concludes: "It's good to be a link in a chain."[24] Che Guevara's definition of the revolutionary *cadre* shows him to be the perfect organizational middleman,

> able to interpret the larger directives emanating from the central authority, make them his own, and convey them as an orientation to the masses . . . someone of ideological and administrative discipline . . . [who] knows how to practice the principle of collective discussion and individual decision making responsibility . . . who can think for himself, which enables him to make necessary decisions and to exercise creative initiation in a way that does not conflict with discipline.[25]

As for the *jeune cadre,* his very social prominence in the early 1960s was the sign of a successful postwar transformation in French corporate bureaucracies away from older patriarchal or heavy-handed styles toward a more flexible, American-inflected practice of management. Modernizing the economic apparatus, as Boltanski has shown, was not a purely technical exercise. It depended first on the stabilization of the greater social order (essentially, containing the rise of the French Communist Party, especially after the wave of strikes in 1947–48), on the creation of a corporate "climate" similar to that of American companies, and thus on a transformation in the *mentalité* of the individual engineer/ executive. In all three levels of operation, the Americans lent a hand. The CIA infiltrated, financed, and organized a split in the CGT after the strikes in the late 1940s, substantially undercutting worker organization and political effectiveness. American-style "rational" modes of

organization and the importation from the United States of psycho-
social technologies such as "human engineering" helped create both a
"climate of confidence" and its new agent, the *jeune cadre*. Charles
Kindleberger, in an influential 1963 study of French economy that included a subsection entitled "The New Men," concluded that the basic
change in the French economy "is one of people and attitudes"; the "new
men" who had taken command were those with expansionist ideas, new
attitudes toward consumption, and a universal belief in the desirability
of growth.[26] Within the corporation, the *cadre* was the intermediary and
the arbitrator standing between employers and the working class; as
such, he was, ideally, the artisan of a new kind of social collaboration
and corporate cohesion. As Daniel Mothé put it, "the *cadre* is he who
defends both the interests of the factory and the interests of the work
force. . . . The *cadre* is thus boss, worker and technician at once."[27] The
importance of status and self-definition through consumption for the
cadre derives from his intermediate position: *cadres* "control" or manage
the workers but are themselves subject to wage contracts as well. Status
helps them define their autonomy.

Such an ambiguous position in the corporate structure—Kracauer
called it "ideological homelessness"—required a sustained level of mental
flexibility and psychological adaptability on the part of the *jeune cadre*.
It is worth noting in passing a singular commentary on the *jeune cadre*
and the new modes of organization that came from an unexpected
source: the films of Jacques Tati. Businessmen figure in all three of his
major films: in *Les vacances de M. Hulot ("Mr. Hulot's Holiday")* (1953)
the radio's background noise provides a constant barrage of business
news and advertisements for such products as "The Briefcase That Spells
Success"; Hulot plays with the little boy whose businessman father is
too busy receiving important international phone calls to pay attention
to him. This last scene is but a brief interlude in *Mr. Hulot's Holiday;* it
is expanded to provide the entire set of dramatic relations in *Mon oncle*
(1956): the little boy who prefers his eccentric, playful uncle (Hulot) to

FIGURE 4.1 *Playtime*

his factory-owner father. In *Playtime* (1967) such easy oppositions have
been rendered moot by accelerated modernization: *vieux Paris* has dis-
appeared and the whole movie transpires in the vague, airport-condo
terrain of prefab industrial space. Whereas the Hulot character functions
overtly as the "other" to the corporate, organization man (at least in the
first two films), he is also just as much (if not more so) his double,
substituting physical flexibility—the elastic body of the clown—for the
mental flexibility in the face of new situations expected of the *cadre*. A
comparison with the clown from an earlier stage in modernization,
Chaplin in *Modern Times,* makes this clear. Tati himself, anxious to
differentiate his work from that of Chaplin, pointed out the main difference between the two clowns: in none of the situations he finds
himself in does Hulot, unlike Charlot, take the initiative. Taking as his
example a scene from *Mr. Hulot's Holiday* where Hulot inadvertently
gets caught up in a funeral procession, Tati comments,

> Take, for instance, the scene in the cemetery, the wreath with
> the dead leaves. Hulot just wanted to take out his car tyre,
> and without his doing anything about it, the leaves stick to
> it and it makes a wreath. If this had happened to Charlot he
> would have deliberately put the leaves on the tyre, in order
> to transform it into a wreath and thus be able to leave the
> cemetery decently. Hulot does not get out, he stays until the
> end, shaking hands with everyone.[28]

Chaplin's clown, in other words, operating at an earlier historical moment in the twentieth century, was still able to exert a kind of energy
or agency, inventing his gags and ultimately triumphing singlehandedly
over the mechanical world whose forces he has unleashed. Hulot does
not unleash the forces that buffet him along, but when they arise, he
reacts with burlesque malleability. He adjusts flexibly to any circum-
stance and, unlike Charlot, never shows himself to be aware that his

actions might seem the least bit funny to other people. This last quality, as Roy Armes points out, allows Hulot to exist on the same level as the other characters in the film in a way that Charlot—singular and heroic—does not. Indeed, by the time of *Playtime* Hulot as a central character has devolved completely, replaced by a series of realistically observed vignettes that he merely passes through. The progressive "leveling" of Hulot goes hand in hand with the leveling of the sharp oppositions in the earlier films between "traditional" and "modern" Paris; the increasing "incorporation" of the character of Hulot into the film *Playtime* is a narrative reflection of the increased standardization of daily life in France.

The identity of both the *jeune cadre* and the revolutionary *cadre,* then, is inseparable from the organization to which each belongs. It is around the set of consumption practices known as "lifestyle," however, that the revolutionary *cadre* and the *jeune cadre* regain their more familiar opposition to each other. The lifestyle of the revolutionary *cadre* or *maquisard,* depicted in novels like *Les enfants du nouveau monde,* in films like *The Battle of Algiers,* in countless campaign diaries, and theorized in a range of texts from the writings of Ho Chi Minh, Che Guevara's diaries and his "Socialism and Man in Cuba," to Mao's "On Sacrifice," is necessarily ascetic, stripped down, and unencumbered. For Guevara, the guerilla was defined by an ethics of sacrifice designed to promote solidarity: "He must join to all these qualities a great spirit of sacrifice, not only for the grand heroic days, but at every moment. To sacrifice oneself in order to help a comrade in little tasks, so that he can do his work, or finish his studies at school, so that he can perfect himself in one way or another. To be always attentive to those around you."[29] Man is defined in the totality of his social relations, and collective discipline is indispensable to the development of his personality. Guerilla testimonios tell the tale of a gradual abandonment of possessions on the mountainside under the indiction to travel light and to "fight self." The loss of possessions is compensated for by a gain in physical strength; physical education, as Mao never ceased pointing out, was the necessary foundation for revolutionary regeneration. To hone and temper the physique through a program of spartan, ascetic training was *part and parcel of* strengthening correct revolutionary sentiments, controlling and mastering the passions. Both Djébar and Pontecorvo in *The Battle of Algiers* represent the Algerian revolutionaries' continuing battle against alcohol and prostitution. The difficulty of such a transformation, and the need for a regime that combined tension with relaxation, can be gauged by the dictum attributed to Ho Chi Minh that "to straighten a curved piece of bamboo, one must bend it in the opposite direction, holding it in that position for a while. Then, when the hand is removed, it will slowly straighten itself."[30] Traveling light is of course also freeing oneself of ideological baggage—the latter facilitated by the revolutionary gender division of labor whereby the women, left behind in the villages as Djébar's novel recounts, become the repositories for traditional memory and continuity, while the men, temporarily freed of that weight and now virilely homeless, are able violently to force the new.

The *jeune cadre,* on the other hand, certainly travels lighter and works harder than his nineteenth-century counterpart, the *honnête bourgeois.* After all, the nineteenth-century bourgeois was defined above all by the simple fact of not working, whereas the *jeune cadre* works harder than anyone else. Nor is the *cadre* bound, for example, to the Sunday table every week for a five-hour gastronomic ritual *en famille* as his predecessor was; he prefers instead lighter and quicker fare. (*L'Express* instituted ready-made "tray" lunches for its employees, the design borrowed from airplane meals.) But his leisure time is largely taken up, as we have seen, with frenetic actual or imagined acquisition and accumulation, and with a kind of cult of comfort centered on the home. Virile asceticism plays no role in the image of the *jeune cadre.* (In fact, if we are to believe Christiane Rochefort, the qualities required of the new middle-class businessman—a certain amorphous adaptability bordering on passivity, serviceability, a pleasant nature, and being on the whole devoid of singularity—amounted to a distinct loss in virility.) The gap separating the two *cadres* need not be geographic at this historical moment: French *cadres* could still be found in the (ex)colonies and revolutionaries in French cities. Their difference is in fact best illustrated by the paradigmatic relation of two fictional couples to the urban space of Paris. For Jérôme and Sylvie in *Les choses,* adrift in their endless window-shopping *flâneries* around the city, "the whole of Paris was a perpetual temptation" (p. 28). For FLN organizer Arezki and his lover Elise in *Elise ou la vraie vie,* on the other hand, driven to walking the streets at night despite the curfew because they have nowhere else to be together, and limited to only a very few areas deemed to be "safe," "Paris was an enormous ambush through which we moved with ludicrous precautions" (p. 188).

IMMOBILE TIME

By comparison the *jeune cadre* and "structural man" had far more in common. Both could walk freely after dark in the streets of Paris during the Algerian War without fear of arrest or police intimidation. Both were members of the new middle class, a class with progressively easier access to credit, and a class to whose pleasures French consumption and urban space were more and more successfully organized and oriented. Both read the same magazines: *L'Express,* whose readership in its initial phase in the early 1950s was primarily made up of "classical" intellectuals—teachers, students, certain of the liberal professions—became, especially after its transformation into a "news magazine," the magazine of the *jeunes cadres* as well, featuring articles on management technique in the same issue with articles about the latest developments within structuralism.[31] And both, it seems, spoke the same language: the hygienic language of technique and efficiency.

For Henri Lefebvre (the most articulate contemporary critic of structuralism), the two men were virtually interchangeable. Structuralism was nothing more than the infusion of technocratic thought into the intellectual field. The structuralist crisis of "man" and "humanism" was above all a practical and historical crisis brought on by a capitalist society where unchecked bureaucratic growth meant that institutions—medicine, teaching, research—no longer put humans first. In a society where objects were more important than people, where cars determine the way people live, why wouldn't the status of "man" be undermined? But instead of analyzing (much less proposing a direction for transforming) that society, structuralism served as an underlying ideology *justifying* the devaluation of humans under capitalist modernization. Rather than theorizing the liquidation of the historical, structuralism enacted and legitimated that liquidation. After all, structuralism's concern was the ordering of objects, not the criticism of their function. The idea that society was composed of agentless structures helped reinforce people's growing sense that the future was not in their control, or that it would play itself out as a kind of slow petrification, that their life was defined by lifeless, meaningless, and unchanging bureaucratic structures ruled by no one. Castoriadis, another contemporary critic, put it this way: "[Structuralism] is the discourse that accords an absolute primacy to science when people are more and more oppressed in the name of science; it wants to persuade them that they are nothing and that science is everything."[32] Structuralism was the ideological handmaiden to the social caste or class represented by the *jeune cadre:* its ideological legitimation, its intellectual veneer.

Lefebvre argued this position in different ways, at times approaching the problem by beginning with the term then dominating contemporary political discourse: technocracy.[33] The word *technocracy* was first created in the United States immediately after World War I to designate a system of organizing economic life inspired by the rational structures of the physical sciences. In France the term was used loosely to designate the post–World War II "reign of the experts": the caste of educated specialists in charge of the big decisions, the *sujets supposés savoir,* the highest level of *cadres,* the so-called *grands corps.* The ascendancy of the technocrats, whose administrative competence was intended to cushion the State from some of the failings of the political parties, in fact began with Vichy. It received a substantial boost under Mendès-France and the crowd of young people who wanted France to "take off," and was especially visible after 1958 when de Gaulle consolidated his return to power surrounded by, for the first time, an elite and overt entourage of *ministres-techniciens*. For Lefebvre technocracy was less harmful in its actions than in the image of itself that it projected to society. That image or ideology—the supreme reign of mature rationality—was not simply an apotheosis of the cult of efficiency, the promise of a maximum of results for a minimum of effort. The ideology of technocracy as mature rationality divides the world into two: those "in the know" are adults, and everyone else is a child. It thus defines all social opposition as "immature" and assumes political immaturity on the part of the Third World. With time, it comes to define social opposition as a largely artificial phenomenon. Social contradiction becomes a thing of the past as an ever-increasing efficiency resolves all social difficulties by augmenting the mass of disposable goods and by distributing them equitably, aided by optimal organization provided by information technology. Even if technocrats systematically favor some group or social system, they invariably place that preferential action under the sign of *consensus*. And consensus, as we know, is reached through rational agreement. At its highest, most evolved degree, the myth of supreme, mature rationality begins to take on the allure of a philosophy: the philosophy of structuralism.

Structural man for Lefebvre was thus above all an impostor. Posing as a philosopher, he offered nothing in the way of a theoretical thought or a discourse critical of his society. Structuralism acted as though it were a theoretical discourse about the society, a knowledge of the social, when in fact it was nothing other than the discourse *of* that society, the chitchat of technocracy, its continuation and ideological reinforcement.[34] Structural man was not a philosopher but a technician, preoccupied, like the *planificateurs* in *Les stances à Sophie,* with breaking apart and recomposing functional systems with the ultimate purpose of maintaining a structure of equilibrium. Lévi-Strauss, in fact, agreed that structural man was not a philosopher. Instead of the image of the technician, however, he preferred to embrace a more *passéiste,* humble, and congenial image, most frequently that of the *bricoleur,* but at times that of the *bricoleur's* close relative, the artisan: "[Structuralism] keeps itself from wanting to found . . . a philosophy. Rather, we consider ourselves to be laboring artisans, bent over phenomena too slight to excite the human passions, but whose value comes from the possibility that, having been seized at that level, they might one day become the object of a rigorous knowledge."[35] But Lefebvre would point out that the very appeal to *rigor* (Lefebvre called it *rigor mortis*) on the part of the structuralists—their continuous embrace of a rhetoric of rigor that would extend well beyond the structuralist galaxy into the various poststructuralist sciences of "reading" like those of Paul de Man—had itself been borrowed from the technocrats so that there might appear to be a rational identity between the object and the act of knowledge at all times.[36] Lefebvre saw structuralism as a sociological phenomenon—that is, as a particular *logic* of the *social* at a specific moment in the history of capitalist modernization, a moment marked by political immobilism and the consolidation of systems. The critique of structuralism is only an episode in the radical critique of capitalist modernization.

The Sartrian battle against structuralism overlapped considerably with that of Lefebvre. Both posited philosophy and history as the two casualties of structuralism, as well as the two sites from which any oppositional critique—like their own—would have to be generated. For Sartre as for Lefebvre there was a direct correspondence between the advent of the technostructure of a new industrial State, the demise of philosophy, and the success of an antihistorical doctrine that negates the subject: "In a technocratic civilization, there is no longer room for philosophy, unless of course it transforms itself into technique. Look at what happened in the United States: philosophy was replaced by the social sciences."[37] Most of the structuralists then beginning to surpass Sartre in intellectual prestige (Lacan, Althusser, Bourdieu, Lévi-Strauss) had, like Sartre, come from the philosophical discipline; their renegade status vis-à-vis their training must have seemed to him to constitute a disciplinary as well as a political betrayal. And where the Sartrian philosophy of language focused on language's endpoint or goal ("The goal [*fin*] of language is to communicate"), structuralism made of communication a general ambiance and language an end in and of itself—measurable in countable units, phonemes and morphemes. But the rise of structuralism in the 1950s and 1960s was above all a frontal attack on historical thought in general and Marxist dialectical analysis in particular; its appeal to many leftist French intellectuals after 1956 was overdetermined by the crisis within the French Communist Party and Marxism following the revelations of Stalin's crimes and the Soviet invasion of Hungary at the end of that year. After such messy historical events, the clean, scientific precision of structuralism offered a kind of respite. One critic describing the sequence from Marxism to structuralism after 1956 put the emphasis on the functionalism of structuralism, and in so doing brought to the forefront the necessary relationship between hygiene and modernization: "It was a kind of ceremonial massacre. . . . It allowed a good mopping-up, a sweep of the broom, a deep breath of air, a hygienic act. You can't always choose the odor of the deodorant or of the cleaning products which is often disgusting, but it does the job."[38]

The vicissitudes of the intellectual career of Roland Barthes during the 1950s is a case in point. What that career shows, perhaps better than any other, is what Marxists and structuralists shared at that moment: a concern with things. But where Marx, in his theory of commodity fetishism, made objects themselves a point of theory or critique, structuralism turned toward taxonomies, toward classifying and organizing the world and its objects. Though never a Marxist, Barthes's *Mythologies* (1957) was written, as he later put it, under the triple aegis of Sartre, Marx, and Brecht.[39] In retrospect, *Mythologies* breaks down into two different books: in it, a long concluding methodological essay, "Le mythe aujourd'hui" written after Barthes's discovery of Saussure, attempts to provide a general theory for the short journalistic pieces he had written and published previously in magazines in the early part of the decade and that make up the body of the book. These short essays were in fact responses to contemporary events (political declarations, expositions, *faits divers*) and key figures (Bardot, Poujade) of the early 1950s, and, most memorably, to the arrival of the reigning objects, the cars and laundry detergents, of the new consumerism. The pleasure of the book comes from the laconic brevity of the essays and their messy contiguity, the jumble of things, people, and events, Greta Garbo next to greasy french fries. *Mythologies*'s intellectual affinities, Sartre, Marx, and Brecht, make sense if we consider the first half of the book at least, the journalistic pieces, to be an exercise in *historic realism*. The Sartrian concept of "situation" dominated the postwar intellectual scene, and the analysis of gestures, acts, objects, and texts in *Mythologies* can be read as a related attempt at "situated" knowledge, a construction of a common social and historical situation. Brecht's role as the champion of the possibility of a vital, contemporary realism was solidified as much by his plays as by the positions he took in his debates with Lukács; Barthes's own interest in Marx was mediated by Brecht to the point where it could be labeled a kind of Brechto-Marxism. As for Marx, he is perhaps best understood, as Derek Sayer points out, as himself a realist, that is, as someone who recognized the limitations of any general theory or model when it comes to analyzing particular historical events, processes, or societies.[40] Material production, for Marx as for Barthes in *Mythologies,* provides the guiding framework within which he then works empirically; empirical phenomena are the starting points of any analysis.

Barthes himself seems to have embraced the realist stance; the final sentence of the book is a call for a kind of realist project of reconciliation between language and its object: "And yet, this is what we must seek: a reconciliation between reality and men, between description and explanation, between object and knowledge" (p. 159). Later he would describe the book as an effort in "social mythology"[41] that had as its object "France itself [*la France réelle*]," a project akin to that of Michelet: "What pleased him in Michelet is the foundation of an ethnology of France, the desire and the skill of questioning historically—i.e., *relatively*—those objects supposedly the most natural: face, food, clothes, complexion. . . . In his *Mythologies,* it is France itself which is ethnographed."[42] He then goes on to establish an affinity between his book and the great realist cosmogonies of the nineteenth century: "Further, he has always loved the great novelistic cosmogonies (Balzac, Zola, Proust), so close to little societies. This is because the ethnological book has all the powers of the beloved book: it is an encyclopedia, noting and classifying all of reality, even the most trivial, the most sensual aspects. . . . Finally, of all learned discourse, the ethnological seems to come closest to a Fiction."[43] A trio of new affinities or identifications have been added to the original trio of Sartre, Brecht, and Marx: here, the author of *Mythologies* is at once a would-be historian (Michelet), a realist novelist (Balzac), and an ethnographer. All three, it seems, aspire to a common goal: the sociohistorical analysis of the object world. Only a few years later, having fallen into what he called "the euphoric dream of scientificity,"[44] Barthes would abandon nonverbal materials and limit himself to analyzing only "language objects" in his *Eléments de sémiologie* (1964). And with his book on fashion, *Le système de la mode* (1967), Barthes renounced the object "clothing" to treat instead only "written clothing," having decided to follow some private advice from that other great ethnographer of the era, Claude Lévi-Strauss.[45]

But Barthes might also have felt pressure to abandon the object world because of the criticism he received for the theoretical essay that concludes *Mythologies,* "Le mythe aujourd'hui." Read today, that essay appears as a kind of hybrid, transitional document. As Philippe Roger points out, it is neither here nor there: too linguistic for the mythological intuitions that precede it and which it attempts to explain; too open to the nonverbal—objects and images—to satisfy linguists.

Immediately after the publication of *Mythologies* in 1957 Barthes began to take on the role of championing the work of Robbe-Grillet and the New Novel in a series of essays published along with his first studies of structuralism. What is striking about the Robbe-Grillet essays, taken as a group, is the degree of identificatory zealotry Barthes brings to the task of supporting the New Novel's attempt to "clean up" novelistic form. The New Novel, for Barthes, was successful only to the extent that it hygienically purged itself of a lengthening list of dark enemies—themes, characters, motivations, plot, diachronic agency. Thus Robbe-Grillet's *Voyeur* receives high praise from Barthes for having "disinfected" or "sterilized [*aseptisé*] the very form of narrative."[46] Barthes's purificatory zeal seems to have surprised even Robbe-Grillet himself, who was no stranger to the urge to cleanse: "Barthes," he wrote in the 1980s after Barthes's death, "only needed my writing as a mopping-up operation [*comme entreprise de nettoyage*]."[47] In effect Robbe-Grillet accuses his most enthusiastic critic of *dirigisme,* of holding the New Novel to a standard of rigorous negation that no literature could ever meet, for his own psychic/intellectual/ethical purposes.

Mopping up from what? If Robbe-Grillet's remarks are accurate, then Barthes's celebration of the New Novel (but celebration connotes too much pleasure; let's say rather his ethical advocacy of the New Novel) corresponds to a kind of ascetic cleansing, a retreat from the pleasures of greasy french fries as much as from the messiness of the real. The vulgar objects of the petit-bourgeois "real" described in *Mythologies* offered Barthes the various satisfactions of slumming: they were both pleasurable and messy; in the end, they made him a bit queasy. The elements that make up *Mythologies* were "everything that provokes in me the nausea of the average, the middle of the road, vulgarity, mediocrity and above all the world of stereotype."[48] The turn to the New Novel and structuralism is a retreat from vulgarity, a situating of the activity of critique under the auspices of scientific rigor rather than those of history, an identification with the engineer rather than the novelist or historian.[49] And rigor is always an ascetic economy, the careful choice of the one, right technique, the precise method exercized under the exacting eye of the Master. But even under the regime of rigor, the old pleasures (the "loved" books by Balzac, the sensuous, material feel of things) creep back; from all the pages of New Novelist prose Barthes read in the early 1960s, what seems to stay with him, almost nostalgically, are the objects, the eyeglasses, erasers, coffeemakers, prefab sandwiches, and cigarettes.[50] Objects "cleansed of all human significance," of course, but objects in fact not so distant from the objects that fill the pages of *Mythologies*. The long list of things that fill these essays acts as an unconscious reminder of Barthes's own perception in *Mythologies* of the "trace" that even the cleanest, most streamlined of objects must bear: "One could answer with Marx that the most natural object contains a political trace, however faint and diluted, the more or less memorable presence of the human act which has produced, fitted up, used, subjected, or rejected it" (pp. 143–144).

As an attack on historical thought itself, we would expect structuralism's effects to be most strongly felt within the field of historiography, on the way history was written during those years and the debates that engaged historians. The very function of historical discourse as interrelation between past and future becomes destabilized during this period. In order to account for this, we must look not only at the history of historiography and the fate of historical analysis at the hands of technique, but we must also consider the French institutions that organize and certify intellectual disciplines, and that determine the hierarchy of disciplines. And we cannot ignore the larger Cold War American hegemony Sartre alludes to when he seems to refer to the United States almost as a laboratory exhibiting life forms into which the French have entered whether they like it or not.

The most immediate context surrounding the structuralist conquest of the humanities was the precipitous and spectacular consolidation of a quantitatively based set of social sciences, as well as the institutions needed to house and implement them, in a few short years immediately after the war. The year 1946 saw the creation of a *Centre d'études sociologiques* (CES) in Paris with a journal to accompany it, the *Cahiers internationaux de sociologie*. Scholars and *planificateurs* alike could benefit from a number of exciting developments, such as the availability of a new and reliable statistical apparatus the groundwork of which had been laid by Vichy: the *Institut national de la statistique et des études économiques* (INSEE), which was established in April of 1946. The newly inaugurated science of statistics was joined by its twin, the science of demographics. The latter was given an institutional home, the *Institut national d'études demographiques* (INED), founded in October 1945 to study the mechanisms of French demographic renewal, and a journal as well: *Population,* which began publication the following year. At the same time the ONI (*L'office national d'immigration*), symbolically situated under the dual leadership of the Ministry of Labor and the Ministry of Population, was created to assure that the immigration made necessary by population loss during the war and by the new industrial priorities' need of a work force be organized in function of the new science of demography.[51] The discipline of psychology won its independence in the university in 1947 by obtaining a specific *"license d'enseignement."* But perhaps the most important institutional birth of the immediate postwar years was that of the *Ecole nationale d'administration* also in 1947. Students graduating from this institution—*énarques,* as they were called—went on to become the new techno-elite of top civil servants and national administrators, the *cadres superieurs.* The prestige of the ENA would rival and eventually surpass that of the "classical" institution whose business had been the forming of the national elite: the *Ecole normale supérieure.* If I place the founding of the ENA in a list of institutions pertaining to the implantation of the social sciences in France it is because the two were adjuncts to each other: state *planification* needed and made direct and frequent use of the knowledge provided by the social sciences.[52] And the new kind of intellectuals trained in the new research institutes and university departments open to demography, economics, and social psychology emerged ready to translate their knowledge immediately to enterprises in the public and private sectors, frequently under contract.

The link between the goals of modernization and the knowledge provided by the social sciences can be gauged by measuring the surge in the latter's institutional backing and prestige in the decade that followed. This was particularly the case with sociology, a field heavily influenced by the methods and language of its American counterpart. In 1955 there were only 20 research centers in the social sciences in France; by 1965 there were more than 300. In 1958, under the direction of Raymond Aron, sociology was granted complete certification by the university in the form of a license and the establishment of a "doctorat de troisième cycle" in sociology. The number of researchers in sociology at the CNRS went from fifty-six in 1960 to ninety in 1964.[53]

The French social sciences we are familiar with now were thus a postwar invention, and like all aspects of French modernization after the war their ascendancy bore some relation to U.S. economic intervention. To a certain extent the turn to this kind of study was funded and facilitated by the United States in a kind of Marshall Plan for intellectuals. A review of the literature makes a convincing case that the foremost American export of the period was not Coca-Cola or movies but the supremacy of the social sciences. In October 1946, the director of the social science division of the Rockefeller Foundation proclaimed, "A New France, a new society is rising up from the ruins of the Occupation; the best of its efforts is magnificent, but the problems are staggering. In France, the issue of the conflict or the adaptation between communism and western democracy appears in its most acute form. France is its battlefield or laboratory."[54] By expanding the social sciences in Europe, Americans sought to contain the progress of Marxism in the world; a science of empirical and quantitative sociology—the study of repetition—was erected against the science of history, the study of event. A grant from the Rockefeller Foundation in 1947 helped finance the founding of the VI section of the *Ecole pratique des hautes études* under the directorship of historian Lucien Febvre, who had seized the initiative from a rival group of sociologists headed by Georges Gurvitch. Home to François Furet in the early 1960s, this institution would be central to the future of the social sciences in France: in 1962, when Febvre's successor Fernand Braudel gathered all the various research *laboratoires* scattered around the Latin Quarter and housed them in a single building on the Boulevard Raspail, the *Maison des sciences de l'homme,* the Ford Foundation helped finance the operation. In 1975 the VI section would in turn emancipate itself from the *Ecole pratique* and become the *Ecole des hautes études en sciences sociales,* with university status and the authorization to grant degrees.[55]

That the names of Febvre and Braudel, two of the most prominent French historians of the twentieth century, should appear as instrumentally active on the side of the social sciences bears explanation. We could call it the sign of the changes going on within the field of history itself as it sought to retain its prewar supremacy against the onslaught of the quantitative sciences. The main tactic employed by the group of historians that came to be known as the Annales school against the threat of structuralism was that of cannibalism: encompass and absorb the enemies as a means of controlling them. Immediately after the war the official journal of the Annales school underwent a significant name change: from *Annales d'histoire économique et sociale* to *Annales: économies, sociétés, civilisations.* Founded in 1929, the original journal had already been concerned with advancing a broad, interdisciplinary notion of historical studies. But the disappearance of the word "history" from the revised title showed a new degree of willingness to embrace the other social sciences, particularly those of demography and economics. Behind the gesture of self-effacement, however, lurked a continuing will to prevail as absent or invisible master.

In a famous article entitled "History and the Social Sciences: the *Longue Durée*," published in *Annales* in 1958, Fernand Braudel responded to the direct attack Lévi-Strauss had leveled against the science of history. Historians, for Lévi-Strauss, were limited in their material to the merely observable; historical knowledge was empirical and thus incapable of arriving at the "deep structures" governing society. Lévi-Strauss, on the other hand, claimed to have determined the way to uncover the fundamental mode of functioning of the human spirit, an invariable that underlay all the surface diversity of human cultures.

Braudel introduces his polemical article by setting up the battle between history, "the least structured of all the human sciences," and its social science neighbors, neighbors that he admits have taught history a great deal.[56] He detects the glimmer of a "common market" on the horizon that would join together all the different fields under a shared project. But by the end of the article it is perfectly clear that such a common market is of interest to Braudel only if history, "the dialectic of duration," can remain king of the enterprise: "For nothing is more important, nothing comes closer to the crux of social reality than this living, intimate, infinitely repeated opposition between the instant of time and that time which flows only slowly. Whether it is a question of the past or of the present, a clear awareness of this plurality of social time is indispensable to the communal methodology of the human sciences" (p. 26). But continuing dominance comes at a stiff price. To remain central, history must, in effect, transform itself and be reorganized as "immobile time."

In the body of the article Braudel presents a critique of the history of events (*l'histoire évenementielle*) and distinguishes instead three distinct historical temporalities: the time of the event (told in "the headlong, dramatic, breathless rush of . . . narrative"); that of the conjuncture (slower-moving economic and social cycles); and that of structures (the long, or very long, *durée*). It is this last, defined as "an organization, a coherent and fairly fixed series of relationships between realities and social masses," that Braudel offers up to Lévi-Strauss. Openly dismissive of Georges Gurvitch, the leading sociologist of the day, and of sociology's attachment to "current events," Braudel is far more respectful and gingerly with an enemy he views as his equal (he refers to Lévi-Strauss as "our guide"). By reorganizing history to be the study of the temporality of structures (the *longue durée*), Braudel incorporates Lévi-Strauss into a larger project under his—Braudel's and history's—direction. The social sciences come together in a "common market" with a shared project that has the *longue durée* at its conceptual center. And since it is still a question—however minimally—of *durée,* of temporal periodization, history can continue to reign supreme.

The far-reaching effects of such an alliance on debates within French historiography are still being felt today. In the 1950s and 1960s Braudel, Le Roy Ladurie, and others, ensconced after 1962 in the *Maison des sciences de l'homme,* produced what Braudel called "a history whose passage is almost imperceptible . . . a history in which all change is slow, a history of constant repetition, ever recurring cycles."[57] Their most formidable enemies within the field of history lived across the street: the long lineage of Marxist historians of the French revolution—Georges Lefebvre, Albert Soboul, and the like—housed at the Sorbonne. For what is at stake in the erasure of the study of social movement in favor of that of structures is the possibility of abrupt change or mutation in history: the idea of Revolution itself. The old-fashioned historians of the event *par excellence* of French history, each one in turn occupying the chaired professorship for the study of the French Revolution instituted by the Sorbonne after 1891, looked askance at their thoroughly modernized, well-funded, and well-equipped (with photocopiers and computers) colleagues across the way.[58]

Despite their equipment, the Annales-school historians had little to say about modernity and nothing at all to say about the modernization they, and to varying degrees the rest of France, were experiencing. They preferred instead the narrative history of feudalism and the *ancien régime* as it relates to structural levels of economy, material civilization, and *mentalités*. In some ways their continued focus on premodern European culture served the same purpose as Lévi-Strauss's concentration on the "dead" or dying cultures of the Amerindians—the purpose of alibi for not dealing with the present, the here and now, or with the conditions that had led to this immediate present. (A similar alibi is provided by Foucault's notion of genealogy: Foucault states that he wants to write about the present; this desire is used to justify the genealogical project, and then he never quite gets to talking about the present.) But it is also possible that historians were projecting their own "eventless" temporality, the steady timeless unfolding of modernization as a naturalized process, back onto the rhythms of the past. Certainly the all-powerful system of connections and communication they discovered in the feudal past under the name of *longue durée* bore some relation to the temporality and conditions of production of the historians themselves. Wasn't the self-reproducing nature of contemporary techno-structures, their almost isomorphic replication, reappearing in the guise of the geographic crystallizations of the distant past? Weren't the historians' experience of the systemic spatial connections of their own time—*"posthistoire"*—determining their formulation and narration of premodern civilization? After two world wars and a messy decolonization, events were to be distrusted. *Longue durée* provided a means of projecting powerlessness, a way of displacing causal agency onto something so vast in scale as to render human action trivial; *longue durée* enshrined the glacial. Change, when it takes place, occurs in a geological timeframe.

I have argued that the major intellectual productions of the 1950s and 1960s, structuralism and Annales-school historiography, cannot be separated from the ideology of capitalist modernization, an ideology that seeks above all to undermine eventfulness by masking the social contradictions that engender events. Annales-school history, through its focus on *mentalité* or shared culture, manifested a powerful disposition toward demonstrating consensuality; it directed its search toward regularities and repetitions, toward establishing the average and the immobile. Just as modernization needed a new recounting of history that would dissolve beginning and end into a natural, quasi-immobile, "spatialized" process, a kind of succession of immobilities, so it needed a system of signs that would establish a common intellectual currency, so to speak, between the various intellectual disciplines, a network of "communication" by which to trace the buried structures common to all social life. In one sense the embrace by the humanities, and by literary studies in particular, of structuralist methodologies was an attempt to wage a battle for continued supremacy in precisely the same way Braudel had waged battle with Lévi-Strauss: by incorporating the enemy. Perhaps by modernizing, the French humanities—history, literature, and above all, philosophy—could retain their classical role in the formation of the country's elites, a role now seriously challenged by the ascendancy of the *Ecole nationale d'administration* and its new technological elite, over the *Ecole normale supérieure*. The "turn to linguistics" offered a way of rationalizing the humanities, rescuing them from archaic isolation and from the provinciality of, for example, the dusty, and narrowly nationalist, chronologies of French literary histories; it offered a conception of literary studies as something more than mere connoisseurship. The linguistic turn put the humanities in communication with other, now more dynamic, disciplines. Scientific methods brought order, establishing the basis for systematic reflection and hygienic pedagogical transmission. Research was now eminently "transcodable": when everything is language, then each of the units in the field of knowledge becomes interchangeable, convertible, and fungible, in a vast, functionalist network. Or, as Mme. Arpel in Tati's *Mon oncle* was fond of exclaiming to the visitors being shown through the connecting spaces of her ultramodern suburban home, "Everything communicates!"

Such an association between Tati's critique of the kind of life it is possible to live inside hypermodern suburban architecture and the contemporaneous reorganization of French intellectual life is less surprising if we remember that the particularity of both Mme. Arpel's home and structuralism is the existence of each as an assemblage of standardized, interconnected components. To create the "total ambiance" of the modernized home, the totality of use-values in the home had to be newly adapted to capitalist mass production through the development of "design": a functionalist aesthetics that would render the components uniform, or compatible, the stove-sink-refrigerator flowing together in a seamless, white-and-chrome unit, the design of the objects and their context mutually reinforcing each other in a seamless ambiance. Similarly, what came to be known as structuralism, and the science of signs it engendered, was less important as a development within any one discipline—anthropology, film studies, literature—than in the communication or transmissions it established among knowledge clusters, institutions, disciplines, and contents of consciousness. To use Braudel's phrase, it established a new interdisciplinary leveling, or common market, of knowledge. Rather than being considered an end or a goal, as in the Sartrian pronouncement that the goal of language is to communicate, "communication" became instead something akin to an ambiance, the milieu for life in the sense that water has for a fish. An ideology "without enemies," an alternative to political ideologies, "communication" announced the installation, thanks to technology, of a kind of neutral, consensual norm in social relations.

But a second fantasy governs the smooth operation of Mme. Arpel's techno-home, and that is its completeness: nothing is left out. This is the reason for Mme. Arpel's compulsive flurry of hygienic activity at moments of entry into or exit from the home: dusting the briefcase, the car bumpers, and so forth. The *cadre* aspires to totality in his or her environment; advertisements from the period excite that desire by promising to supply what they call *"le luxe intégrale."* Functionalist luxury: a

FIGURE 4.2 *Mon oncle*

FIGURE 4.3 *Godin* appliances advertisement, *Elle*, October 1954.

total system made up of the set of integers that, taken individually, mean very little but together equal perfection. A kind of paranoiac construction—like all utopias—the *cadre*'s home works by inversion. If nothing is missing, then nothing extraneous—no threats to security—will enter. The seamlessness and impermeability of the *cadre*'s home (most effectively rendered in *Mon oncle* in the hilarious scene where Mme. Arpel mistakes her exotically dressed neighbor ringing at the gate for a rug salesman and sends her away, and in all of the numerous gags concerning the gate as instrument of exclusion) bears a relation to the totalizing breadth to which structuralism aspired: the dream of regrouping all the humanities and social sciences around the study of the sign. Nothing is missing, so nothing extraneous—history, context, event, the nondiscursive in short—can enter.

Nothing is missing so nothing *is* extraneous. Like the automobile, structuralism's initial appeal was the lure of wide-open spaces, of an expansive opening out onto other disciplines, other knowledges. And just as mass access to the automobile ultimately meant the monadic, cramped experience of the traffic jam, so the very breadth of structuralism's reach ended up in a kind of paranoiac cul-de-sac: in its denial of history as the realm of the unexpected and the uncertain, its denial of the extradiscursive, its denial, finally, of the outside itself. In 1958 at the peak of structuralism's hegemony, that outside took several forms. It was what lay outside Mme. Arpel's electronic fortress: the nonsynchronous and the uneven, the lived experience of a whole range of French people who, just across the empty lots, from the point of view of the modernized, "lag behind," "drag us down," and are lacking in standards of hygiene. But the outside was above all the experience of those who had taken history into their own hands, the various dark "new men" being formed in the crucible of anticolonial struggle—people who at that moment embodied real political contradiction and thus real alterity to "the stasis of Europe" and the smooth, steady functioning of ahistorical structural systems.

This is the peculiar view that this conjuncture produced, a view that slammed shut the Algeria chapter and relegated it to another temporality, made it and all of colonialism into an instant archaism. It was a view associated with the triumph of what Jacques Marseille and others have termed "the modernist movement," that is, a renovated and improved French capitalism: "the coming together of a group of men for whom the analysis of numerical figures for the most part obliterated political passion, for whom the power of France was thus not identified with the possession of an empire—the vestige of a Malthusian past that they condemned."[59] This perspective dominated the French lurch into modernization, and it is essential glue to the current social formation in France today. A dominant contemporary French perspective holds its colonial past to be an "exterior" experience, added on but not essential to French historical identity—an episode that ended, cleanly, in 1962. France closed that chapter and moved on to bigger autoroutes, all-electric kitchens, and the European Economic Community. But as Etienne Balibar suggests, even the economic and cultural Americanization of France could not have been felt as a metaphorical colonization if it did not intervene on the persistent base of France's own constitutive identity as a colonizer.[60] France's denial of the ways in which it was and is formed by colonialism, its insistence on separating itself off from what it views as an extraneous period irrelevant to its true national heritage, forms the basis of the neoracist consensus of today: the logic of segregation and expulsion that governs questions of immigration, and attitudes toward immigrants, in France. The logic of exclusion has its origins in the ideology of capitalist modernization, an ideology that presents the West as a model of completion, thus relegating the contingent and the accidental—the historical, in a word—to the exterior. To understand contemporary rightist slogans such as "France must not become Africa," we must return to the era of "Algeria *is* France," and to the time when France believed wholeheartedly in the possibility of limitless and even development.

NOTES

INTRODUCTION

1

Françoise Giroud, *Leçons particulières* (Paris: Livres de poche, 1990), p. 123. Here, and throughout this book, translations from the French are mine unless otherwise noted.

2

See Alain Robbe-Grillet, *Pour un nouveau roman* (Paris: Editions de Minuit, 1963), trans. Richard Howard as *For a New Novel* (New York: Grove, 1965).

3

Francis-Louis Closon, *Un homme nouveau: L'ingénieur économiste* (Paris: Presses universitaires françaises, 1961), pp. 13–14.

4

Alain Touraine, *La société post-industrielle* (Paris: Editions Denoël, 1969), pp. 115–116.

5

Michel Aglietta, *Régulation et crises du capitalisme* (Paris: Calmann-Lévy, 1976), trans. David Fernbach as *A Theory of Capitalist Regulation: The U.S. Experience* (London: New Left Books, 1976), p. 159.

6

Arguments, an important neo-Marxist journal published between 1956 and 1962 for which Lefebvre, Morin, and Barthes, among others, wrote, published a translation of Georg Lukács's essay on reification in 1960.

7

André Gauron, *Histoire économique de la Vième république,* vol. 1 (Paris: Maspero, 1983), p. 6.

8

See Alain Lipietz, "Governing the Economy in the Face of International Challenge: From National Developmentalism to National Crisis" in *Searching for the New France,* ed. James F. Hollifield and George Ross, (New York: Routledge, 1991), pp. 17–42.

9

Richard Kuisel's recent *Seducing the French: The Dilemma of Americanization* (Berkeley: University of California Press, 1993) is a good example of a political/economic history that focuses entirely on the "French economic miracle" and Americanization without any consideration of the end of empire.

10

The phrase is André Gorz's. See his *Critique de la division du travail* (Paris: Seuil, 1973).

11

Cited in Louis Chevalier, *L'assassinat de Paris* (Paris: Calmann-Lévy, 1977), trans. David P. Jordan as *The Assassination of Paris* (Chicago: University of Chicago Press, 1994), p. 236.

12

See Etienne Balibar, "L'avancée du racisme en France," in *Les frontières de la démocratie* (Paris: La Découverte, 1992), pp. 19–98.

13

I have made this argument in the context of a reading of the detective fiction of Didier Daeninckx in "Watching the Detectives," in *Postmodernism and the Rereading of Modernity,* ed. Francis Barker, Peter Hulme, and Margaret Iversen, (Manchester: Manchester University Press, 1992), pp. 46–65.

Chapter 1

1

Quoted in Guy Herzlich, "Adieu Billancourt," *Le Monde* (March 29, 1992), p. 25.

2

Emile Pouget, quoted in Anson Rabinbach, *The Human Motor* (New York: Basic Books, 1990), p. 241.

3

See Pierre Naville, *L'état entrepreneur: Le cas de la Régie Renault* (Paris: Anthropos, 1971).

4

See Simone Weil, *La condition ouvrière* (Paris: Gallimard, 1951) and Robert Durand, *La lutte des travailleurs de chez Renault racontée par eux-mêmes, 1912–1944* (Paris: Editions sociales, 1971); more recent examples of the genre include Daniel Mothé, *Militant chez Renault* (Paris: Seuil, 1965), and Robert Linhart, *L'établi* (Paris: Editions de Minuit, 1978), trans. Margaret Crosland as *The Assembly Line* (Amherst: University of Massachusetts Press, 1981). Claire Etcherelli's novel *Elise ou la vraie vie* (Paris: Editions Denoël, 1967), trans. June Wilson and Walter Benn Michaels as *Elise or the Real Life* (New York: Morrow, 1969), is at least in part a fictionalized testimonio, since Etcherelli worked in car factories.

5

See *René-Jacques,* edited by Pierre Borhan and Patrick Roegiers (Paris: Editions La Manufacture, 1991).

6

See Belden Field, "French Maoism," in *The 60s without Apology,* ed. Sohnya Sayers, Anders Stephanson, Stanley Aronowitz, and Fredric Jameson (Minneapolis: University of Minnesota Press, 1984), pp. 148–177.

7

Linhart, *L'établi,* pp. 13–14.

8

See Henri Lefebvre, *La vie quotidienne dans le monde moderne* (Paris: Gallimard, 1968), p. 104. The English translation by Sacha Rabinovitch, *Everyday Life in the Modern World* (New York: Harper, 1971), renders this term as "compulsive time" (p. 53).

9

Françoise Sagan, *Avec mon meilleur souvenir* (Paris: Gallimard, 1984); trans. Christine Donougher as *With Fondest Regards* (New York: Dutton, 1985), p. 65.

10

Jean Baudrillard, *Le système des objets* (Paris: Gallimard, 1968), p. 94.

11

Françoise Sagan, *Réponses* (Paris: Editions Jean-Jacques Pauvert, 1974), trans. by David Macey as *Réponses: The Autobiography of Françoise Sagan* (Godalming, England: The Ram Publishing Company, 1979), p. 57. Louis Chevalier dates the "undrivability" of Paris a bit earlier: "It seems to me that around 1953–54, maybe a bit earlier, people began to speak more and more frequently, more and more strongly, more and more forthrightly, in a voice carrying real authority, as they say, of how difficult it was to drive in Paris" (*The Assassination of Paris,* trans. David P. Jordan [Chicago: University of Chicago Press, 1994], p. 52). Chevalier is referring to a popular discourse—the need to ease traffic congestion—that helped pave the way for the redevelopment of the city throughout the 1950s and 1960s designed to accommodate Paris to the automobile.

12

See Dominique Borne, *Petits bourgeois en révolte? Le mouvement poujade* (Paris: Flammarion, 1977): "In the expansion [of Poujadism], the car played an essential role. The movement's propaganda disseminated countless images of Pierre Poujade's *tours de France,* carefully counting up the kilometers. The automobile opened up the possibility of being in one town to oppose an inspection, and then in another hundreds of miles away a few hours later. When a department was "invested," Pierre Poujade and his crew were everywhere . . . the movement was propagated like a rumor: 'It was the wind,' Pierre Poujade told us in 1975" (p. 26). Poujade's followers paid for a new car for him by subscription after his old one was destroyed in a wreck. See Stanley Hoffmann, *Le mouvement poujade* (Paris: Armand Colin, 1956), p. 244.

13

Jules Romains, "Introduction," in *L'automobile en France* (Paris: Régie nationale des usines Renault, 1951), pp. 20–21.

14

See Richard Kuisel, *Capitalism and the State in Modern France* (Cambridge: Cambridge University Press, 1981), pp. 263–267.

15

Quoted in Jacques Borgé and Nicolas Viasnoff, *La 4CV* (Paris: Balland, 1976), p. 78.

16

Ibid., p. 65.

17

Ibid., p. 78.

18

See, for example, Romains, "Introduction," p. 4: "I've always had a lively attachment to that machine so 'friendly to man' [*amie de l'homme*] that is the auto."

19

Roland Barthes, "La voiture, projection de l'égo," *Réalités* 213 (1963).

20

Guy Debord, "Theses on Traffic," in *Situationist International Anthology,* ed. and trans. Ken Knabb (Berkeley: Bureau of Public Secrets, 1981), p. 56. See also Roland Barthes, *Mythologies* (Paris: Seuil, 1957), translated by Annette Lavers as *Mythologies* (New York: Noonday, 1972); Baudrillard, *Le système des objets;* Alain Touraine, *L'évolution du travail ouvrier aux usines Renault* (Paris: Centre national de la recherche scientifique, 1955).

21

Cited in Marc Dambre, *Roger Nimier: Hussard du demi-siècle* (Paris: Flammarion, 1989), p. 554.

22

See Jean-Pierre Bardon, *La révolution automobile* (Paris: Albin Michel, 1977): one in ten French owned a car in 1960; one in four in 1972 (p. 224), and "the relative gap between the degree of motorization of the least motorized categories (agricultural workers, service personnel, manual workers and salaried workers) and those in the most motorized categories (management, liberal professions, industrialists and large shopowners) was maintained between 1953 and 1969" (p. 235).

23

See Victoria de Grazia, "Mass Culture and Sovereignty: The American Challenge to European Cinemas, 1920–1960," *Journal of Modern History* 61 (March 1989), pp. 53–87.

24

Ibid., p. 56.

25

Under the terms of this contract France would get $650 million in international credit, and a $720 million allowance on American surplus; in addition, the United States canceled an outstanding lend-lease bill of nearly 2 billion dollars. In return the French agreed to abandon the price-equalization procedures and the protective tariffs of the prewar era, and, in particular, "to admit American motion pictures into France for the first time since 1939."

26

Dennis Turner, "Made in the USA: Transformation of Genre in the Films of François Truffaut and Jean-Luc Godard," Ph.D. dissertation, University of Indiana, 1981, p. 3.

27

See Wolfgang Schivelbusch, *The Railway Journey: Trains and Travel in the 19th Century* (New York: Urizen, 1979).

28

Georges Perec, *Les choses* (Paris: René Julliard, 1965); trans. David Bellos as *Things* (Boston: Godine, 1990), p. 55.

29

Karl Marx, *Grundrisse: Foundations of the Critique of Political Economy,* trans. Martin Nicolaus (London: New Left Review, 1973), p. 534. Italics in the original.

30

Jean-Luc Godard, *Jean-Luc Godard par Jean-Luc Godard* (Paris: Editions Pierre Belfond, 1968), p. 383.

31

Nimier, known as the "French James Dean," was said to have introduced himself to Gaston Gallimard with the words "I have come, my dear sir, to change ink into gasoline." Gallimard later made him a present of an Aston-Martin that Nimier christened the "Gaston-Martin." See Dambre, *Roger Nimier,* p. 515.

32

Ibid., p. 463.

33

Ibid., p. 552.

34

See Jacques Becker, "Enquette sur Hollywood," *Cahiers du cinéma* 54 (December 1955), p. 73.

35

Jean-Luc Godard, *Introduction à une véritable histoire du cinéma* vol. 1 (Paris: Editions Albatros, 1980), p. 92.

36

Jean Narboni and Tom Milne, eds., *Godard on Godard* (New York: Viking, 1972), p. 183.

37

See, for example, François Truffaut, "Feu James Dean," *Arts* (September 26, 1956); see also the April 27, 1957, article in *Le monde* by Georges Hourdin, "Le malheur et la fureur de vivre," which establishes a comparison between Dean and Françoise Sagan in part by way of their respective car accidents: "The novelist and the actor translated each other, in their art, in their attitude toward life—what we've all taken to calling *le mal de la jeunesse*." In January 1959 *Combat* published a portrait of Roger Nimier called "Un James Dean français." Novelist Jean-René Huguenin, who would die at the wheel in 1962 a week before Nimier, and filmmaker Eric Rohmer published articles on Dean in the late 1950s in *Arts*. At least two French books on Dean were published in 1957 alone: Yves Salgues's *James Dean ou le mal de vivre* and Raymond de Becker's *James Dean ou l'aliénation insignifiante*. Both were savagely reviewed by François Truffaut in *Arts* (April 24, 1957).

38

Carné's choice of St-Germain-des-Prés as the nightlife setting of choice for his jaded youth is not surprising: the golden *quartier* of the immediate postwar period, center to an intellectual and cultural life based on the presence of publishing houses rather than universities, it was one of the first areas targeted for massive gentrification, already underway by the mid-1950s.

39

See Hervé Hamon and Patrick Rotman, *Génération: Les années de rêve* (Paris: Seuil, 1987), pp. 49–51.

40

Jacques Demy's *Lola* (1960) opts for a similarly optimistic narrative of reconciliation. Through most of this movie Lola, a bar girl played by Anouk Aimée, vacillates in a desultory way between two lovers: Cassard, whose Frenchness is marked by his tired warmed-over existentialist rhetoric of ennui and liberty, and Frankie, a good-natured American sailor. Her dilemma is resolved with the reappearance of her first love, Michel, the father of her child, who returns in a giant white convertible (the "voiture de rêve") and whisks her away. With Michel, Lola can have the best of both worlds, French and American: rich like an American, with an American's accoutrements (car, cowboy hat), he is still the "first love," the authentic one in relation to whom all subsequent lovers have been pale, unsatisfactory reproductions, the one who speaks an unaccented "mother tongue."

41

Norma Evenson, *Paris: A Century of Change, 1878–1978* (New Haven: Yale University Press, 1979), pp. 54–55.

42

Ibid., p. 58.

43

Quoted in Pierre Lavadon, *Nouvelle histoire de Paris: Histoire de l'urbanisme à Paris* (Paris: Hachette, 1975), p. 536.

44

Françoise Sagan, *Réponses,* p. 121.

45

Marc Angenot, *1889: Un état du discours social* (Quebec: Editions du préambule, 1989), p. 332.

46

Chevalier, *The Assassination of Paris,* p. 49.

47

By 1962, 840,000 copies of *Bonjour tristesse* had been sold in France, and 4,500,000 copies in translation.

48

On the vicious debate surrounding mass-market paperback editions of "classics," see François Maspero, "Livres de poche et culture de masse," *Partisans* 16 (June–August 1964), pp. 65–70; and the special issue of *Les temps modernes* devoted to the debate (April 1965).

49

Françoise Sagan, *Bonjour tristesse* (Paris: Julliard, 1954); trans. Irene Ash as *Bonjour tristesse* (New York: Dell, 1955), pp. 52–53, translation altered.

50

Françoise Sagan, *Aimez-vous Brahms . . .* (Paris: René Julliard, 1959); trans. Peter Wiles as *Aimez-vous Brahms . . .* (New York: Dutton, 1960), p. 11, translation altered.

51

Simone de Beauvoir, *Tout compte fait* (Paris: Gallimard, 1972), trans. Patrick O'Brian as *All Said and Done* (New York: Putnam, 1974), p. 122: "In this it was a question of making the silence speak—a new problem for me."

52

Henri Lefebvre, *Le temps des méprises* (Paris: Stock, 1975), p. 34.

53

Perec, quoted in David Bellos's introduction to the American edition of *Things*, trans. David Bellos (Boston: Godine, 1990), p. 9.

54

Beauvoir, *All Said and Done*, p. 122.

55

Christiane Rochefort, *Les stances à Sophie* (Paris: Grasset, 1963), p. 68.

56

Christiane Rochefort, *Les petits enfants du siècle* (Paris: Grasset, 1961), p. 84.

57

Luc Boltanski, "Accidents d'automobile et lutte des classes," *Actes de la recherche en sciences sociales* (March 1973), p. 31. See also his "Les usages sociaux de l'automobile: Concurrence pour l'espace et accidents," *Actes de la recherche en sciences sociales* (March 1975), pp. 25–49. Boltanski's studies set out to analyze what other sociologists were content merely to register: the staggering number of road deaths in France, which by the mid-1960s had surpassed those of all the other countries at the same socio-economic level. See Jacques Vallin and Jean-Claude Chesnais, "Les accidents de la route en France. Mortalité et morbidité depuis 1953," *Population* 3 (May–June 1975), pp. 443–478. France-at-the-wheel had become something of an international carica-ture, particularly after a series of accidents involving well-known people; the decade of the sixties is inaugurated by the car wreck that killed Albert Camus and Michel Gallimard on January 5, 1960. On Mondays the daily newspapers used a black-bordered square to set off the death toll of the preceding weekend, "still more murderous than the one before" (Hamon and Rotman, *Génération*, p. 291). But this did not prevent dailies such as *Le Monde* from adopting a lurid, frankly fascinated tone in their reporting of particular accidents that lent prestige to the pulverized machines: "The powerful automobile was going very fast—130 kilometers an hour according to some—when suddenly it lurched to the side of the road, though the road was perfectly straight . . ."; "He was driving along when suddenly, for no apparent reason, the car made a terrible swerve . . ." *Le monde* of the early 1960s, cited in Jean-Pierre Quélin, "Vitesse Grand V," *Le monde* (August 24, 1991), p. 10.

58

Simone de Beauvoir, entretien avec Jacqueline Piatier, *Le monde,* Sélection hebdomadaire (December 29–January 4, 1966).

59

Simone de Beauvoir, *Les belles images* (Paris: Gallimard, 1966), trans. Patrick O'Brian as *Les belles images* (New York: Putnam, 1968), p. 28. I have adapted O'Brian's translation in some instances.

60

Beauvoir, *All Said and Done,* p. 122.

61

Highly visual passages such as these recall certain driving scenes from *Bonjour tristesse* and other films of the period, suggesting another complicity between driving and the movies: together they conspired to make the late 1950s the "age of the profile."

62

Deirdre Bair, *Simone de Beauvoir* (New York: Simon and Schuster, 1990), pp. 431–432.

63

Jean-Jacques Servan-Schreiber, *Le défi américain* (Paris: Editions Denoël, 1967), trans. Ronald Steel as *The American Challenge* (New York: Atheneum, 1968), p. 257.

64

Françoise Giroud, *Leçons particulières,* (Paris: Livres de poche, 1990), p. 149.

65

Sartre and Beauvoir had themselves upstaged the previously reigning cultural and political couple, Elsa Triolet and Louis Aragon, immediately after the war.

66

Françoise Giroud, *Françoise Giroud vous présente le tout-Paris* (Paris: Gallimard, 1952), p. 298.

67

Servan-Schreiber's brother, Jean-Louis, writes in his memoir, "It was natural to identify with John Kennedy, who embodied the leading ideas of my generation:

America, youth, success, beauty, the future." *A mi-vie: L'entrée en quarantaine* (Paris: Stock, 1977), p. 137.

68

Françoise Giroud, *Leçons particulières,* pp. 164–165.

69

Ibid., p. 172.

70

Françoise Giroud, *Si je mens . . .* (Paris: Stock, 1972); trans. Richard Seaver as *I Give You My Word* (Boston: Houghton Mifflin, 1974), pp. 130–131, translation altered. Evidence of Giroud's obsession with the figure of Beauvoir can be found throughout her autobiographical writing. During an interview, for example, when accused of being incapable of admitting she has made a mistake, Giroud replies, "Who do you take me for? Simone de Beauvoir?" (p. 204). In a later text she feels called upon to situate her views on feminism in relation to Beauvoir's:

> It seemed to me that such a long history as the history of the domination of women by men demanded a more subtle analysis than the one written by Simone de Beauvoir. Surely many women will become conscious, thanks to her, of their possibilities, of the pressures that have forced them off the track. But at the same time I didn't feel concerned by *Le deuxième sexe.* The way I felt myself to be a woman had nothing to do with a superstructure, a placement that had been imposed on me to the point of deforming me. It was just the opposite.
>
> The feminine part of me was the essential, the fundamental, the skeleton on which everything else hung (*Leçons particulières,* pp. 126–127).

71

Giroud, returning to the offices of *L'Express* after an absence of several years, is "struck by the closed shutters and the electric lighting in the middle of the day, by the decoration in a hyperamerican style as well" (Serge Siritzky and Françoise Roth, *Le roman de l'Express 1953–1978* [Paris: Atelier Marcel Jullian, 1979], p. 340); "the atmosphere was no longer that of a large family" (p. 331).

72

Siritzky and Roth, *Le Roman de l'Express,* p. 328.

CHAPTER 2

1

Alphonse Boudard, *La fermeture* (Paris: Editions Robert Laffont, 1986), p. 14.

2

Françoise Giroud, *I Give You My Word,* trans. Richard Seaver (Boston: Houghton Mifflin, 1974), p. 108.

3

François Maspero, *Les passagers du Roissy-Express* (Paris: Seuil, 1990), pp. 171–172.

4

Barthes, *Mythologies,* trans. Richard Howard as *The Eiffel Tower* (New York: Noonday, 1979), p. 49.

5

See Jean Baudrillard, *Le système des objets* (Paris: Gallimard, 1968), pp. 249–252. Baudrillard analyzes a Pax ad in such a way as to argue that advertising omits any representation of the real contradictions of society in favor of an imaginary creation of a presumed collectivity: "The example of Pax is clear: advertising tries to create a solidarity between individuals on the basis of a product the purchase and usage of which is precisely what sends each person back to his own individual sphere" (p. 251). Collective nostalgia, or rather nostalgia for some lost, imaginary collectivity, serves to fuel individual competition and to further or enable what Lefebvre, Castoriadus, and others were calling "privatization."

6

Boudard, *La fermeture,* p. 16.

7

Robert Paxton, *Vichy France: Old Guard and New Order, 1940–1944* (New York: Columbia University Press, 1972), pp. 330–331.

8

Mme. Richard, responsible in large part for the law of April 13, 1946, that closed French brothels, was not alone in her quest. Raymond Bossus of the French Communist Party agreed: "Paris must maintain its status as the leading world capital and not allow itself to be dirtied any more." Cited in Boudard, *La fermeture,* p. 45.

9

Alain Robbe-Grillet, *Pour un nouveau roman,* (Paris: Editions de Minuit, 1963); trans. Richard Howard as *For a New Novel* (New York: Grove, 1965), p. 9.

10

See Jacques Leenhardt's *Lecture politique du roman* (Paris: Editions de Minuit, 1973); see also Fredric Jameson, "Modernism and Its Repressed," in *The Ideologies of Theory,* vol. 1 (Minneapolis: University of Minnesota Press, 1988), pp. 167–180.

11

See Giroud, *I Give You My Word,* pp. 127–128. Giroud's survey revealed that 25 percent of French women never brushed their teeth, and that 39 percent washed themselves once a month.

12

Frantz Fanon, "L'Algérie se dévoile," in *L'an V de la révolution algérienne* (Paris: Maspero, 1959), trans. Haakon Chevalier as "Algeria Unveiled," in *A Dying Colonialism* (New York: Grove, 1965): "This enabled the colonial administration to define a precise political doctrine: "If we want to destroy the structure of Algerian society, its capacity for resistance, we must first of all conquer the women.'" (pp. 37–38).

13

Georges Perec, *Les choses* (Paris: René Julliard, 1965); trans. David Bellos as *Things* (Boston: Godine, 1990), p. 24.

14

In 1961, for example, 755,000 copies of each issue of *Elle* were sold; 1,132,000 copies of *Marie-Claire.* See Evelyne Sullerot, *La presse feminine* (Paris: Armand Colin, 1983), p. 83. Readership of magazines is more difficult to gauge, since each copy of a magazine is commonly read by more than one person. According to *Elle* magazine itself, one out of every six French women read *Elle* in 1955.

15

The first postwar issue of *Marie-Claire* sold out after only a few hours on the newsstands; the next issue reported 500,000 copies sold.

16

Françoise Giroud, *Leçons particulières* (Paris: Livres de poche, 1990), p. 122.

17

Giroud, *I Give You My Word,* pp. 106–108.

18

See Henri Lefebvre, *Critique de la vie quotidienne,* vol. 2 (Paris: Arche, 1961), pp. 84–91, for his discussion of women's press.

19

See especially, in addition to Barthes's *Mythologies* and Lefebvre's *Critique de la vie quotidienne,* vol. 2, Edgar Morin, *L'esprit du temps* (Paris: Grasset, 1962).

20

One of the most interesting of Grégoire's discoveries comes from her comparison between French and American magazines of the period regarding how sexuality is treated. American magazines offer comparatively far more explicit information on female sexual pleasure and on reproduction and abortion issues than the French magazines that, Grégoire argues, are governed by provincial mores and must avoid shock in the countryside. See Ménie Grégoire, "La presse feminine," *Esprit* (July–August 1959), pp. 17–34.

21

Françoise Giroud, "Apprenez la politique," *Elle* (May 2, 1955).

22

Marcelle Segal, one of the most popular advisors to the lovelorn *(Courrier du coeur),* quoted in Grégoire, "La presse feminine," p. 26.

23

Serge Siritzky and Françoise Roth, *Le roman de l'Express 1953–1978* (Paris: Atelier Marcel Jullian, 1979), p. 202.

24

See Françoise Sagan, "La jeune fille et la grandeur," *L'express* (June 16, 1960); reprinted in Simone de Beauvoir and Gisèle Hamini, *Djamila Boupacha* (Paris: Gallimard, 1962); trans. Peter Green (New York: Macmillan, 1962), pp. 245–246.

25

Jean-Jacques Servan-Schreiber, quoted in Michel Winock, *Chronique des années soixante* (Paris: Seuil, 1987), p. 66.

26

Grégoire, *"La presse feminine"* p. 24.

27

Emile Zola, *Au bonheur des dames* (Paris: Gallimard, 1980), trans. anonymously as *The Ladies' Paradise,* intro. Kristin Ross (Berkeley: University of California Press, 1992), p. 93.

28

Anon., "La propreté de l'enfant," *Marie-Claire* (May 1955), pp. 98–99.

29

Jean-Pierre Rioux, *La France de la Quatrième République,* vols. 1 and 2 (Paris: Seuil, 1980–1983), trans. Godfrey Rogers as *The Fourth Republic 1944–1958* (London: Cambridge University Press, 1987), p. 370.

30

Dominique Ceccaldi, *Politique française de la famille* (Paris: Privat, 1957), quoted in Claire Duchen, "Occupation Housewife: The Domestic Ideal in 1950s France," *French Cultural Studies* 2 (1991), p. 4.

31

Parisian fascination with *les arts ménagers* dates from the 1920s: the first of the yearly expositions opened to 100,000 visitors in 1923. In 1926 the *Salon des arts ménagers* took up a more luxurious residence in the Grand Palais. The *Salon* was not held from 1939 to 1948; when it resumed, attendance mushroomed, and by 1955 there were about 1.5 million visitors. See Yvette Lebrigand, "Les Archives du *Salon des arts ménagers,*" *Bulletin de l'institut d'histoire du temps présent* (December 1986), pp. 9–13.

32

1953 Salon des arts ménagers, Actualités Gaumont.

33

Lefebvre, *Critique de la vie quotidienne,* vol. 2, p. 88.

34

Elsa Triolet, *Roses à crédit,* vol. 1 of *L'age du nylon* (Paris: Gallimard, 1959), p. 31.

35

See Edgar Morin, *Commune en France: La métamorphose de Plodémet* (Paris: Fayard, 1967). Morin's study is the best of the many ethnographies of rural France that appeared in the 1960s as intellectuals began to chart the final days of the peasant class in France. Whereas books such as Eugen Weber's *Peasants into Frenchmen* and Beauroy's *The Wolf and the Lamb* had told the story of the passing of the old peasant culture and the arrival of a new urban mass culture, focusing on the period 1870–

1914 as the moment when the old folk culture finally died, Morin argues that the old culture survived into the 1950s and was killed off only by consumer culture and mass media. The *exode rural* became heavy in the 1950s, with between 100,000 and 150,000 people leaving the countryside for the cities each year. (Rioux, *The Fourth Republic,* p. 181). Morin argues that peasant women were the secret agents of modernization.

36

Actually the key fetish item that begins the process of Martine's transition is a phosphorescent statue of the Virgin Mary that her hairdresser friend brings her from Lourdes. Combined in this small fetish, bequeathed by the good, clean "modern" mother, are all the necessary elements for Martine's transformation: shininess, curative bathing, the sacred, and transcendent femininity.

Martine's choice of occupation reflects a significant social phenomenon: in the six-year period between 1952 and 1958 the number of people employed in hairdressing salons tripled (Rioux, *The Fourth Republic,* p. 329).

37

See Adrian Rifkin, *Street Noises: Parisian Pleasure 1900–1940* (Manchester: Manchester University Press, 1993), p. 66. Rifkin's book poses this question in the context of an illuminating discussion of the life and songs of French realist *chanteuses* compared with the career of Maurice Chevalier.

38

Morin, *L'esprit du temps,* p. 138.

39

Boris Vian, "Complainte du progres," in *Chansons et poèmes* (Paris: Editions Tchou, 1960), pp. 95–98.

40

In making the original boutique one that sold umbrellas, Demy is participating in a time-honored French tradition: that of using the umbrella to register the outmoded or artisanal world in the face of mass production and accelerated commodification. The best example is the closing chapters of Zola's *Au bonheur des dames,* where the character Bourras's handmade umbrella and cane shop becomes the last holdout against the monstrous department store's takeover of the entire *quartier.* See also Louis Aragon's *Le paysan de Paris* (Paris: Gallimard, 1923), for a nostalgic evocation of an umbrella and cane shop: "There was an honorable cane merchant who offered a clientele difficult to satisfy a selection of numerous luxurious articles, crafted in such a way as to please to both the body and the wrist" (pp. 29–33).

41

In 1958 one in ten French households contained a refrigerator; three years later 40 percent had one, and by 1969 the statistic was 75 percent. (Winock, *Chronique des années soixante,* p. 112).

42

See Walter Benjamin on the horror of bourgeois interiors, especially the fragment entitled "Louis-Philippe or the Interior," in *Charles Baudelaire: A Lyric Poet in the Era of High Capitalism,* trans. Harry Zohn (London: New Left Books, 1973), pp. 167–169.

43

Triolet, *Roses à crédit* p. 81.

44

Paulette Bernège, cited in Duchen, "Occupation Housewife" p. 5.

45

See Adrian Forty, *Objects of Desire* (New York: Pantheon, 1986).

46

Christiane Rochefort, *Les stances à Sophie,* (Paris: Grasset, 1963), p. 100.

47

Baudrillard, *Le système des objets,* p. 41.

48

Roland Barthes, "La voiture, projection de l'égo," *Réalités,* 213 (1963), p. 45.

49

See Cornelius Castoriadis, "Le mouvement révolutionnaire sous le capitalisme moderne," *Socialisme ou barbarie* 31–33 (December 1960, April and December 1961); trans. David Ames Curtis as "Modern Capitalism and Revolution" in *Political and Social Writings,* vol. 2 (Minneapolis: University of Minnesota Press, 1988), pp. 226–343.

50

Alain Touraine, *La société post-industrielle* (Paris: Editions Denoël, 1969), p. 78.

51

Elle magazine provided one such study, reporting the results of a survey in 1954 that showed most French people naming their leading (and largely unrealizable) dream to be home ownership. See *Elle* (March 22, 1954), "En 1954 les français font 5 rêves."

But the dream was older than that: immediately after the war the Institut National d'Etudes Démographiques conducted a survey that showed 72 percent of French people saying they wanted a single-family dwelling, with 28 percent willing to add a half-hour commute to get one. See Norma Evenson, *Paris: A Century of Change, 1878–1978* (New Haven: Yale University Press, 1979), p. 251.

52

Lefebvre, *Critique de la vie quotidienne,* vol. 3, pp. 61–62.

53

In Agnes Varda's film *Cléo de 5 à 7,* the same metonymy occurs: a taxi radio droning in the background juxtaposes news of rioting in Algeria with advertisements for a new shampoo *à l'américaine* made of whiskey: "Scotch revitalizes your hair!"

54

Henri Lefebvre, *La somme et le reste* (Paris: Méridiens Klincksieck, 1989), p. 171.

55

Roger Trinquier, *La guerre moderne* (Paris: Editions de la Table Ronde, 1961); trans. Daniel Lee as *Modern Warfare* (New York: Praeger, 1964), p. 29.

56

The extent of French government censorship during these years was unprecedented. Journals such as *L'express* were seized regularly (twelve times between 1958 and 1962 alone). For the most thorough account of the censorship, see Martin Harrison, "Government and Press in France during the Algerian War," *The American Political Science Review* 58, no. 2 (June 1964), pp. 273–285.

57

See the introduction to Henri Alleg, *La question* (Paris: Editions de Minuit, 1961), p. 10.

58

Henri Alleg, *The Question,* trans. John Calder (New York: Braziller, 1958), p. 46.

59

Georges Perec, *Things,* p. 21.

60

Cited in Rita Maran, *Torture: The Role of Ideology in the French-Algerian War* (New York: Praeger, 1989), p. 145.

61

Benaissa Souami, twenty-seven-year-old political science student, cited in Jerome Lindon, ed., *La gangrène* (Paris: Editions de Minuit, 1959); trans. Robert Silvers as *The Gangrene* (New York: Lyle Stuart, 1960), p. 42.

62

Pierre Leulliette, *St. Michel et le Dragon* (Paris: Editions de Minuit, 1961); trans. as *St. Michael and the Dragon* (London: Heinemann, 1964), p. 233.

63

Cited in Xavier Grall, *La génération du Djebel* (Paris: Editions du Cerf, 1962), p. 34.

64

In 1954, the year the Algerian revolution began, 59 percent of French households had running water—up from 37 percent in 1946. By 1968 this figure had reached 90.8 percent. In 1954 28 percent of French households had toilets inside; by 1968 54.8 percent did. In 1954 only 17 percent of Parisians had a shower or bath in their dwelling. If "comfortable" is defined by the trio of central heating, a toilet inside, and a shower or bath, only 6 percent of French households had achieved this in 1954, as opposed to 63 percent in the United States. See Jean Fourastié, *Histoire du comfort* (published in 1950 as *Les Arts ménagers*; reprint, Paris: PUF, 1973), pp. 106–110.

65

Pierre Vidal-Naquet, *La torture dans la république* (Paris: Editions de Minuit, 1972), p. 13.

66

Ian Birchall mentions this in the context of a reading of Etcherelli's *Elise ou la vraie vie;* see his "Imperialism and Class: The French War in Algeria," in *Europe and Its Others,* vol. 2, ed. Francis Barker, Peter Hulme, Margaret Iversen, and Diana Loxley, (Colchester: University of Essex, 1985), pp. 162–174.

67

Maurienne [Jean-Louis Hurst], *Le déserteur* Paris: Editions de Minuit, 1960; reprint, Paris: Editions Manya, 1991, p. 21.

68

See for example, Claude Bourdet, writing in a March 29, 1956, article in *France-Observateur*: "A hundred thousand young Frenchmen are in danger of being thrown away in the dirty war in Algeria."

69

"Civilizing mission" discourse frequently plays a role in torture sessions, as in the testimony of twenty-six-year-old Moussa Khebaili, who reports a French police officer saying, "You're one of a race I hate, like the Negroes. Now you're going to see what France is really made of—you're just a bunch of slaves. We taught you how to shit in a hole" (Lindon, *The Gangrene,* p. 69).

70

Jean-Jacques Servan-Schreiber, *Lieutenant en Algérie* (Paris: René Julliard, 1957), trans. Ronald Matthews as *Lieutenant in Algeria* (New York: Knopf, 1957), p. 3; later in this narrative, a French soldier laments, "The army is the only thing we've got left that's clean" (p. 93).

71

See Maran, *Torture,* pp. 84–85.

72

Cited in the original edition of Vidal-Naquet, *La torture dans la république,* published in English as *Torture: Cancer of Democracy* (Baltimore: Penguin, 1963), p. 137, translation altered.

73

See Jean-Paul Sartre, "Une Victoire," afterword to Alleg, *La question,* pp. 99–122. The most useful philosophical and political meditation on the history of torture is Page duBois's recent *Torture and Truth* (New York: Routledge, 1991).

74

Benaissa Souami, cited in Lindon, *The Gangrene,* p. 46.

75

Etcherelli, *Elise ou la vraie vie,* pp. 137–138, translation altered.

76

Fanon, *A Dying Colonialism,* p. 24.

77

The morality/functionality distinction played an important role in structuring the debate within the French intellectual circle protesting the war. Much of the protest was conducted on moral grounds involving outrage against the French army's use of torture; other intellectuals, such as Simone de Beauvoir, believed with Fanon that such torture was unexceptional, simply the logic of colonialism: "To protest in the name of morality against 'excesses' or 'abuses' is an error which hints at active complicity. There are no 'abuses' or 'excesses' here, simply an all-pervasive *system*." Beauvoir and Hamini, *Djamila Boupacha*, p. 19. See also Jean-Pierre Rioux and Jean-François Sirinelli, eds., *La guerre d'Algérie et les intellectuels français* (Paris: Editions Complexe, 1991), especially the articles by the editors and that of Marie-Christine Granjon, "Raymond Aron, Jean-Paul Sartre et le conflit algérien."

78

Roger Trinquier, cited in Vidal-Naquet, *Torture,* English edition, p. 55.

79

Vidal-Naquet, *Torture,* English edition, p. 40.

80

Bernard Droz and Evelyne Lever, *Histoire de la guerre d'Algérie 1954–1962* (Paris: Seuil, 1982), p. 140.

81

The term turns up constantly in torture testimonios such as Alleg's; in the testimony of Benaissa Souami, for example, we read, "Finally, he got tired, left, and was replaced by three 'specialists,' as they called themselves" (Lindon, *The Gangrene,* p. 36); another Algerian salesman speaks of being put "in condition" before the arrival of the "specialists." (Bechir Boumaza, in Lindon, *The Gangrene,* cited on pp. 75–76).

82

Droz and Lever, *Histoire de la guerre d'Algérie,* p. 140.

83

Ibid., pp. 140–141.

84

Henri Marrou, "France, ma patrie," in *Le monde* (April 5, 1956).

85

Jean-Pierre Vittori, *Nous, les appelés d'Algérie* (Paris: Stock, 1977), pp. 153–154.

86

Vidal-Naquet reports that in an early stage of the revolution *harkis* (Arabs fighting on the side of the French) were made to conduct torture sessions so that French officers could "keep their hands clean" (*Torture,* English edition, p. 44). The *harki*-torturer appears as a major character in Assia Djébar's classic novel about the revolution, *Les enfants du nouveau monde* (Paris: René Julliard, 1962). The French officer's concern for hygiene takes on a darkly comical appearance in a passage from Alleg's narrative:

> I looked at C——, who this time was accompanied by S——. He was in civilian clothes, very elegant. I had to clear my throat, and he stepped away from me, saying: "Look out! He's going to spit!"
>
> "What does it matter to us?" one of the others asked.
>
> "I don't like it, it's not hygienic." He was in a hurry and he was afraid of getting his suit dirty. (Alleg, *The Question*, pp. 74–75).

87

Cited in Benjamin Stora, *La gangrène et l'oubli: La mémoire de la guerre d'Algérie* (Paris: La Découverte, 1991), p. 29.

88

Sorting centers, sometimes translated as "transit centers," were centers for the movement, storage, and processing of human material.

89

Vidal-Naquet, *Torture*, English edition, p. 53.

90

Ibid., p. 56.

91

Attributed most notably to Michel Debré, prime minister under de Gaulle.

92

Fanon, *A Dying Colonialism*, p. 31.

CHAPTER 3

1

"L'Algérie c'est la France," François Mitterand cited in Jean-Pierre Rioux, *The Fourth Republic 1944–1958,* trans. Godfrey Rogers (London, Cambridge University Press, 1987), pp. 238–239.

2

See *Les temps modernes* (November 1955); see also the article by communist geographer Jean Dresch in *La pensée* (July 1956) entitled "Le fait national algérien": "Algeria is not France. . . . But look at the miserable dispute that has arisen surrounding what it should be called instead. Everyone agrees that Algeria has a particular originality. But we are horrified by the idea of nationality, by the fact of the Algerian nation."

3

Cited in Raymond Aron, *France: Steadfast and Changing: The Fourth to the Fifth Republic* (Cambridge: Harvard University Press, 1960), p. 111. Aron, in fact, was one of the first to articulate a right-wing position supporting a negotiated peace in Algeria, having recognized early on that political independence for Algeria was quite compatible with the interests of French imperialism.

4

Slogan disseminated by the *Union pour le salut et le renouveau de l'Algérie française* (USRAF), an association launched in April 1956; cited in Charles-Robert Ageron, "'L'Algérie dernière chance de la puissance française,': Etude d'un mythe politique (1954–1962)," *Relations internationales* 57 (Spring 1989), p. 121.

5

M. E. Naegelen, governor general in Algeria, in a speech to the National Assembly in December 1954, cited in Ageron, "L'Algérie dernière chance de la puissance française," p. 113.

6

Two million posters bearing this slogan were distributed by the *Union pour le salut et le renouveau de l'Algérie française.*

7

Maurice Thorez, cited in Hervé Hamon and Patrick Rotman, *Les porteurs de valises: La résistance française à la guerre d'Algérie* (Paris: Seuil, 1982), pp. 25–26.

8

Edgar Faure, cited in Hamon and Rotman, ibid., p. 30.

9

Editorial published in *Le Monde* (February 28, 1956).

10

This comes from the famous formula of an unnamed French head of state, quoted by Frantz Fanon in *Pour la révolution africaine* (Paris: Maspero, 1964), trans. Haakon Chevalier as *Toward the African Revolution* (New York: Grove, 1967), p. 83.

11

Journalist Roger Moralès writing in the *Presse de l'union française,* cited in Ageron, "L'Algérie dernière chance de la puissance française," p. 116.

12

Cited in Fanon, *Toward the African Revolution,* p. 161.

13

Ibid., p. 102.

14

Cited in Benjamin Stora, *La gangrène et l'oubli: La mémoire de la guerre d'Algérie* (Paris: La Découverte, 1991), p. 18. Stora himself is partial to the marriage metaphor, entitling the first chapter of his book "France, 1954–1962: The Dark Violence of Family Secrets," and a subsequent chapter, "Violent Divorce."

15

The "allocations familiales" in fact dated from the early 1930s when they were instituted to help raise the birth rate after World War I. The post–World War II version carried some innovations: in addition to the government stipend to large families, an "allocation de la mère au foyer" was added in 1955—a direct rewarding of the at-home mother—as was a one-time bonus to mothers who had a baby within the first two years of marriage, the *prime à la naissance.* The opening pages of Rochefort's *Les petits enfants du siècle* recount the narrator's internalized sense of being *maudite,* or literally "belated," in that her birth took place two weeks too late for her parents to receive a *prime.*

16

Edgar Morin, *L'esprit du temps,* (Paris: Grasset, 1962), p. 157. Roland Barthes provided another contemporary analysis of the new domesticity in *Mythologies*: "Happiness, in this universe, is to play at a kind of domestic enclosure: 'psychological' questionnaires, gadgets, puttering, household appliances, schedules, the whole of this utensile paradise of *Elle* or *L'Express* glorifies the closing of the hearth, its slippered introversion, everything which occupies and infantilizes it, excusing it from a broader responsibility: 'two hearts, one hearth.'" *The Eiffel Tower,* trans. Richard Howard (New York: Noonday, 1979), pp. 24–25.

17

Pierre Poujade, cited in Dominique Borne, *Petits bourgeois en révolte? Le mouvement poujade* (Paris: Flammarion, 1977), p. 190.

18

Pierre Poujade, quoted in Stanley Hoffman, *Le mouvement poujade* (Paris: Armand Colin, 1956), p. 211.

19

Cited in Jules Romains, "Introduction," in *L'automobile de France* (Paris: Régie nationale des usines Renault, 1951), p. 7.

20

Françoise Mallet-Joris, *Les signes et les prodiges* (Paris: Grasset, 1966), p. 23.

21

Even Jérôme and Sylvie, the least given to introspection of the characters, are plagued by an occasional "odd, almost worrisome feeling that they were not quite grasping something" (Perec, *Things,* trans. David Bellos [Boston: Godine, 1990], p. 82).

22

Georges Perec, "Georges Perec Owns Up: An Interview with Marcel Benabou and Bruno Marcenac," *The Review of Contemporary Fiction* 13 (Spring 1993), no. 1, p. 18.

23

Derek Sayer, *The Violence of Abstraction* (Oxford: Blackwell, 1987). Sayer is paraphrasing E. P. Thompson, whom he cites: "Class is defined by men as they live their own history and, in the end, this is its only definition" (p. 21).

24

Perec, "Georges Perec Owns Up," p. 17.

25

Pierre Poujade, *J'ai choisi le combat* (Saint-Céré: Société générale des éditions et des publications, (1955), p. 26.

26

See Henri Lefebvre, *La somme et le reste,* (Paris: Méridiens Klincksieck, 1989): "Poujadism . . . is antibureaucracy on the level of the bureaucracy. The small shopkeeper wishes his role as a functionary of the State to be consecrated, that he have the tangible benefits and prestige that derive from this, and at the same time—contradicting himself—he wishes that he be set free, that the State, the machine and the system fall to pieces" (p. 196).

27

See Michel Aglietta, *A Theory of Capitalist Regulation: The U.S. Experience,* trans. David Fernbach (London: New Left Books, 1976), pp. 59–60.

28

See the opening pages of *Les belles images.* The "remodeled farmhouse" [*ferme amenagée*] plays an important role in Martine's marriage in *Roses à crédit:* unable to accept life on her husband's working farm, Martine's excuse is that it doesn't resemble any of the pictures of *fermes amenagées* she has seen in magazines.

29

Georges Perec, cited in Andrew Leak, "Phago-citations: Barthes, Perec, and the Transformation of Literature," *Review of Contemporary Fiction* (Spring 1993), p. 64.

30

Perec, "Georges Perec Owns Up," p. 19.

31

Simone de Beauvoir, *All Said and Done*, trans. Patrick O'Brian (New York: Putnam, 1974), p. 122.

32

Françoise Giroud, *I Give You My Word*, trans. Richard Seaver (Boston: Houghton Mifflin, 1974), p. 127, translation altered. Giroud's phrase in French is "faire 'décoller' la France"; I translate this as "to make France 'take off'" because I think Giroud is referring explicitly to W. W. Rostow's well-known term designating one of the five stages necessary, in his view, for the transition from a traditional society to one of high mass consumption: the "take-off." In his book *The Stages of Economic Growth: A Non-communist Manifesto* (Cambridge: Cambridge University Press, 1960), the bible of capitalist modernization, Rostow describes the take-off as the interval during which the old blocks and resistances to steady growth are finally overcome.

Published in editions of 60,000 in 1953, *L'Express* had reached a circulation of 500,000 in 1967.

33

Luc Boltanski, "America, America . . . le plan Marshall et l'importation du 'management,'" *Actes de la recherche en sciences sociales* no. 38 (1981), p. 25. Mendès-France, the anti-Poujade figure of his age and a Jew, figured prominently in the Poujadist imaginary. Poujade was fond of stating that "our fathers, who went to bistros, were at Verdun and Mèndes wasn't there"; another more succinct Poujadist slogan was "Mendès back to Jerusalem."

34

See Benedict Anderson, *Imagined Communities: Reflections on the Origin and Spread of Nationalism* (London: Verso, 1983).

35

The link between magazine reading and a national imaginary is made clear in a humorous way by Rochefort's entitling the magazine in her novel *"France-Femme"*— a magazine that doesn't in fact exist but that underlines the frequent use of "France" in names: *Marie-France, France-Observateur,* etc. In Mallet-Joris's *Les signes et les prodiges,* which is about the founding of a conservative glossy magazine designed to promote *pied-noir* culture in France in the wake of Algerian independence, characters debate what to name their journal: *La France d'en haut* [France from on High] is rejected for sounding too much like a parish bulletin, *Vraie France* for being too banal; the only point of agreement is that they must have "France" in the title; finally they decide, simply, on *La France.*

36

Along with magazine reading it is radio, still, and not television, that provides a media environment uniting characters. Viewers of the films of Godard, Varga, and Tati, for example, heard for the first time radio advertising, in cars and homes, as continuous background noise.

37

Ernst Bloch, "Nonsynchronism and the Obligation to Its Dialectics," trans. Mark Ritter, *New German Critique* no. 11 (Spring 1977), pp. 22–38.

38

Jérôme and Sylvie's frustrations at the Tunisian bazaars should be juxtaposed with their ecstasies back in Paris at the local flea markets when they would happen upon "the slightly imperfect surplus stock of America's most celebrated shirt-makers" (Perec, *Things,* p. 40). Andrew Leak uses these shirts as an example of Jérôme and Sylvie's embodiment of a "hand-me-down," petit-bourgeois ideology—petit-bourgeois ideology being but a degraded version of bourgeois ideology. Maybe so, but they also show France's "hand-me-down" position vis-à-vis the United States at this time.

39

See Denise Dubois-Jallais, *La tzarine: Hélène Lazareff et l'aventure de 'ELLE'* (Paris: Editions Robert Laffont, 1984), p. 138; and Communica International, *De la 4CV à la vidéo 1953–1983: Ces trentes années qui ont changé notre vie* (Paris: Communica International, 1983), p. 14.

40

Claude Eveno and Pascale de Mezamat, eds., *Paris perdu: Quarante ans de bouleversements de la ville* (Paris: Editions Carré, 1991), p. 275.

41

Norma Evenson, *Paris: A Century of Change, 1878–1978* (New Haven: Yale University Press, 1979) pp. 309–310.

42

Eveno and de Mezamat, *Paris perdu,* p. 159.

43

For more on cultural movement in Paris under the Commune, see Kristin Ross, *The Emergence of Social Space: Rimbaud and the Paris Commune* (Minneapolis: University of Minnesota Press, 1988).

44

Evenson, *Paris,* p. 238.

45

See the section on Algerians in Marianne Amar and Pierre Milza, *L'immigration en France au XX siècle* (Paris: Armand Colin, 1990), pp. 36–44. By metonymy too, the *bidonvilles* were synonomous in the popular French imagery with North Africans when the largest of these, at Champigny-sur-Marne, was actually Portugese. See also Paul E. White, "Immigrants, Immigrant Areas and Immigrant Communities in Postwar Paris," in *Migrants in Modern France,* ed. Philip Ogden and Paul White (London: Unwin Hyman, 1989), pp. 195–212.

46

Amar and Milza, *L'immigration en France,* p. 42.

47

Cited in François Maspero, *Les passagers du Roissy-Express,* (Paris: Seuil, 1990) p. 196.

48

Louis Chevalier, *The Assassination of Paris,* trans. David P. Jordan (Chicago: University of Chicago Press, 1994), p. 260. The final night of operation of Les Halles was February 27, 1969. See Chevalier's discussion of the debates surrounding the removal of the market, pp. 210–216.

49

Chevalier, *The Assassination of Paris,* p. 34. Chevalier cites Fernand Pouillon's history of the vocabulary used to designate the new social type: "Certainly I knew well the profession of 'the real estate businessman,' (the term 'developer' was later invented by Larrue, the director of the Comptoir national for housing, created in 1954–55)" (p. 31).

50

See Siegfried Kracauer, *Orpheus in Paris: Offenbach and the Paris of His Time,* trans. Gwenda David and Eric Mosbacher (New York: Knopf, 1938). The career of Georges Pompidou is a good example of the new ease of transfer between public and private realms effectuated under Gaullism: Pompidou was a member of de Gaulle's cabinet in 1945, entered the Rothschild group in 1954, became director of the bank in 1956, returned to de Gaulle's cabinet (de Gaulle having become once again president) for six months in 1958, took up direction of the bank again in 1959, and became prime minister in 1962. See Eveno and de Mezamat, *Paris perdu,* pp. 159–162.

51

See Bruno Duriez, "De l'insalubrité comme fait politique," *Espaces et sociétés* no. 30–31 (July–December 1979), pp. 37–55.

52

Hygiene and security were dubbed "the two tits of the bourgeoisie" by the authors of the best study of the redevelopment of Belleville. They suggest that Belleville's history as the center of insurgency during the Paris Commune helped contribute to its twentieth-century image as unsafe: the image of worker-anarchists assassinating an archbishop easily merging into that of knife-wielding foreigners. See Jean Céaux, Patrick Mazet, and Tuoi Ngo Hong, "Images et réalités d'un quartier populaire: le cas de Belleville," *Espaces et sociétés* no. 30–31 (July–December 1979), pp. 71–107.

CHAPTER 4

1

Alain Touraine, *La société post-industrielle* (Paris: Denoël, 1969), p. 7.

2

Aimé Césaire, *Discours sur le colonialisme* (Paris: Présence Africaine, 1955); trans. Joan Pinkham as *Discourse on Colonialism* (New York: Monthly Review Press, 1972), p. 9.

3

Frantz Fanon, *Les damnés de la terre* (Paris: Maspero, 1961), trans. Constance Farrington as *The Wretched of the Earth* (New York: Grove, 1968), p. 314.

4

See Mark Poster, *Existential Marxism in Postwar France* (Princeton: Princeton University Press, 1975). Poster quotes Lefebvre's definition of the "total man" (p. 56): "What is the total man? Not physical, physiological, psychological, historical, economic or social exclusively or unilaterally; it is all of these and more, especially the sum of these elements or aspects; it is their unity, their totality, their becoming."

5

Roland Barthes, "L'activité structuraliste," in *Essais critiques* (Paris: Seuil, 1964); trans. Richard Howard as "Structuralist Activity," in *Critical Essays* (Evanston: Northwestern University Press, 1972).

6

Georges Balandier, *Histoire des autres* (Paris: Stock, 1977), p. 187.

7

Michel Crozier, "The Cultural Revolution: Notes on the Changes in the Intellectual Climate in France," *Daedalus* no. 2 (1964), p. 540.

8

See François Dosse, *Histoire du structuralisme,* vol. 1 (Paris: La Découverte, 1991), pp. 324–334.

9

The text of the manifesto read as follows: "1. We respect and judge justified the refusal to take up arms against the Algerian people. 2. We respect and judge justified the conduct of those French who believe it is their duty to lend aid and protection to Algerians oppressed in the name of the French people. 3. The Algerian people's cause, which contributes decisively to the ruin of the colonial system, is the cause of all free men."

Well-known intellectuals who signed the declaration included Sartre, Beauvoir, Marguerite Duras, Maurice Blanchot, Henri Lefebvre, Michel Leiris, Alain Robbe-Grillet, Christiane Rochefort, and Pierre Vidal-Naquet.

10

Claude Levi-Strauss, *La pensée sauvage* (Paris: Librairie Plon, 1962); trans. George Weidenfeld as *The Savage Mind* (Chicago: University of Chicago Press, 1966), p. 247.

11

Michel Foucault, *Les mots et les choses* (Paris: Gallimard, 1966); translated as *The Order of Things* (New York: Random House, 1970), p. 387.

12

In a curious parallel, Poujadist rhetoric also waged battle against the abstraction of "man" by economists and philosophers—in favor, of course, not of some revolutionary collectivity but of the primacy of the individual. Poujadism was a movement based firmly on experience, upon, as its leader put it, "not the 'man' of the philosophers, but living men; not an abstraction of men, but men who are quite real . . . men, good and bad, who laugh and who cry . . ." Pierre Poujade, cited in Sean Fitzgerald, "The Anti-Modern Rhetoric of Le Mouvement Poujade," *Review of Politics* 32 (April 1970), p. 180.

13

Jacques Rancière uses these examples taken from French workers in developing what remains the best critique of Althusser in *La leçon d'Althusser* (Paris: Gallimard, 1974).

14

See Albert Memmi, *Portrait du colonisé précédé du portrait du colonisateur* (Paris: Editions Corréa, 1957); trans. Howard Greenfield as *The Colonizer and the Colonized* (New York: Orion Press, 1965), p. 132.

15

See ibid., p. 86; Fanon, *The Wretched of the Earth*, pp. 42–43; and Césaire, *Discourse on Colonialism*.

16

Frantz Fanon, *A Dying Colonialism*, trans. Haakon Chevalier (New York: Grove, 1965), p. 30, Fanon's emphasis.

17

Ernesto Che Guevara, *Le socialisme et l'homme* (1967; reprint Paris: La Découverte, 1987), pp. 94–108.

18

Jacques Leenhardt, *Lecture politique du roman* (Paris: Editions de Minuit, 1973), p. 165. After independence the *cadre* and the "new middle classes" make an abrupt appearance in African nations as well, in the form of the national bourgeoisie Fanon warned about in "The Pitfalls of National Consciousness" in *The Wretched of the Earth:* a class of educated, government functionaries that come to be designated as an "administrative bourgeoisie." See Immanuel Wallerstein, "La bourgeoisie: concept et réalité du XI au XXI siecle," in *Race, nation, classe,* ed. Etienne Balibar and Immanuel Wallerstein (Paris: La Découverte, 1988); trans. Chris Turner as "The Bourgeois(ie) as Concept and Reality," in *Race, Nation, Class: Ambiguous Identities* (London: Verso, 1991), p. 141.

19

Poujade, cited in Fitzgerald, "The Anti-Modern Rhetoric of Le Mouvement Poujade," pp. 169, 170.

20

Jean Dubois has assembled some of the most commonly used phrases from the period used to qualify the *jeune cadre*: "men of the future," "those responsable for our time," "knights of modern industry," "the new elite on whom we must rely," "the leaders of the new society," "the new aristocrats designated by competence more than by blood or wealth." See his *Les cadres, enjeu politique* (Paris: Seuil, 1971).

For Poujade the *cadre* was the composite image of the enemy; *cadres* were those who manipulate, make plans and projects (*fonctionnaires* and *budgetivores),* who coldly administer, modernize, make files; they were ideologues, those who graduate from the Polytechnic, those who go to school for 25 years, the technocrats and bureaucrats who want to turn the people of 1789 into robots. See Dominique Borne, *Petits bourgeois en révolte? Le mouvement poujade* (Paris: Flammarion, 1977), pp. 181–198.

21

Francis-Louis Closon, *Un homme nouveau: L'ingénieur économiste* (Paris: Presses universitaires françaises, 1961), p. 9.

22

See Luc Boltanski, "America, America . . . le plan Marshall et l'importation du 'management,'" *Actes de la recherche en sciences sociales* no. 38 (1981), or his longer study, *Les cadres: La formation d'un groupe sociale* (Paris: Editions de Minuit, 1982); trans. Arthur Goldhammer as *The Making of a Class: Cadres in French Society* (London: Cambridge University Press, 1987).

23

Closon, *Un homme nouveau,* p. 10.

24

Assia Djébar, *Les enfants du nouveau monde* (Paris: René Julliard, 1962), p. 291.

25

Che Guevara, "Cadres: Backbones of the Revolution," in *Che Guevara and the Cuban Revolution: Writings and Speeches of Ernesto Che Guevara,* ed. David Deutschmann (Sydney: Pathfinder, 1987), pp. 170–171.

26

Charles Kindleberger, "The French Economy," in *In Search of France* (Cambridge: Harvard University Press, 1963), pp. 118–158.

27

Daniel Mothé, *Militant chez Renault* (Paris: Seuil, 1965), p. 88.

28

Jacques Tati, cited in Roy Armes, *French Cinema since 1946,* vol. 1 (New York: A.S. Barnes, 1966), p. 147.

29

Guevara, *Le socialisme et l'homme,* pp, 45–46.

30

Cited in Eric Wolf, *Peasant Wars of the Twentieth Century* (New York: Harper and Row, 1969), p. 192.

31

Pascal Ory and Jean-Francois Sirinelli, *Les intellectuels en France* (Paris: Armand Colin, 1992), pp. 206–207.

32

Cornelius Castoriadis, *La société française* (Paris: 10/18, 1979), p. 226.

33

Lefebvre wrote a series of articles against structuralism throughout the late 1950s and 1960s; these were collected in a book entitled *Au delà du structuralisme* (Paris: Anthropos, 1971), and re-edited under the title of *L'idéologie structuraliste* (Paris: Anthropos, 1971).

34

Lefebvre's critique of structuralism thus departs substantially from that of E. P. Thompson in *The Poverty of Theory* (London: Merlin, 1978). Thompson criticized the embrace of structuralism or "theory" by a heretofore solidly pragmatic British left; Lefebvre, on the other hand, criticizes structuralism for being insufficiently theoretical, for having banished political concepts such as "alienation" necessary for an understanding and transformation of contemporary France.

35

Claude Lévi-Strauss, "Du bon usage du structuralisme," *Le monde* (January 13, 1968).

36

Thus Lévi-Strauss, when questioned about his own engagement with political causes or movements, responded, "No, I think that my intellectual authority, to the extent that I'm recognized as having one, rests on the sum of my work, and on my scruples of rigor and exactitude." *De près et de loin* (Paris: Odile Jacob, 1988), p. 219.

37

Jean Paul Sartre, "Jean-Paul Sartre répond," *L'arc* 30 (1966), p. 94.

38

René Lourau, cited in Dosse, *Histoire du structuralisme,* vol. 1, p. 202.

39

See Roland Barthes, *Roland Barthes* (Paris: Seuil, 1975); trans. Richard Howard as *Roland Barthes* (New York: Hill and Wang, 1977), p. 145.

40

See Derek Sayer, *The Violence of Abstraction* (Oxford: Blackwell, 1987), pp. 12–13 and p. 151n.

41

Barthes, *Roland Barthes,* p. 148.

42

Ibid., p. 84.

43

Ibid., pp. 84–85.

44

Roland Barthes, "Réponses," *Tel quel* 47 (Autumn 1971), p. 97.

45

See Philippe Roger, *Roland Barthes, roman* (Paris: Editions Grasset & Fasquelle, 1986), p. 90.

46

Barthes, *Essais critiques,* p. 70, my translation.

47

Alain Robbe-Grillet, *Le miroir qui revient* (Paris: Editions de Minuit, 1984), p. 69.

48

Barthes, cited in Dosse, *Histoire du structuralisme,* p. 102.

49

The engineer constituted the ideal narrator of the New Novel; in 1958 Claude Ollier's *La mise en scène* (Paris: Editions de Minuit, 1958), featuring a mining engineer under contract to build a network of roads in an obscure mountainous region of North Africa, won the Prix Médicis.

50

See, in particular, "Littérature objective," "Littérature littérale," "Il n'y a pas d'école Robbe-Grillet," and "Le point sur Robbe-Grillet," all in Barthes, *Essais critiques*.

51

Its philosophy, clearly, was that of favoring an immigration *"de peuplement"* of European families deemed culturally assimilable. Algerians escaped the control of this office because they were regarded as legally French. The dream of an influx of European families faded after 1954 when it became clear that the immigration actually occurring was largely that of single North African men.

52

"The *planificateur* asks sociologists to add to economic planning what is missing from it." Claude Gruson, "Planification économique et recherche sociologique," *Revue française de sociologie* 5, no. 4 (October–December 1964).

53

See François Dosse, *L'histoire en miettes: Des "Annales" à la "nouvelle histoire"* (Paris: La Découverte, 1987), pp. 99–101.

54

J. H. Willit, cited in B. Mazon, "Fondations américaines et sciences sociales en France: 1920–1960," thesis, Ecole des hautes études en sciences sociales, 1985, p. 103.

55

See Olivier Betourné and Aglaia Hartig, *Penser l'histoire de la Révolution* (Paris: La Découverte, 1989), pp. 135–136 and 145–146.

56

Fernand Braudel, "Histoire et sciences sociales. La longue durée," *Annales ESC,* no. 17 (December 10, 1958), pp. 725–753; reprinted in *Ecrits sur l'histoire* (Paris: Flammarion, 1969); trans. Sarah Matthews as *On History* (Chicago: University of Chicago Press, 1980).

57

Fernand Braudel, cited in Peter Burke, *The French Historical Revolution: The Annales School 1929–89* (Stanford: Stanford University Press, 1990), p. 36.

58

For an informative history of the battle between the Annales and the Sorbonne, or between Furet and Soboul, see Bétourné and Hartig, *Penser l'histoire de la Révolution.* I thank Linda Orr for drawing my attention to this book.

59

Jacques Marseille, *Empire colonial et capitalisme français (années 1880–années 1950), histoire d'un divorce* (Paris: Albin Michel, 1984), p. 351.

60

See Etienne Balibar, *Les frontières de la démocratie,* (Paris: La Découverte, 1992), pp. 57–65.

SELECTED BIBLIOGRAPHY

Ageron, Charles-Robert, "'L'Algérie dernière chance de la puissance française': Etude d'un mythe politique (1954–1962)," *Relations internationales* 57 (Spring 1989), pp. 113–139.

Aglietta, Michel, *Régulation et crises du capitalisme* (Paris: Calmann-Lévy, 1976); trans. David Fernbach as *A Theory of Capitalist Regulation: The U.S. Experience* (London: New Left Books, 1976).

Alleg, Henri, *La question* 1958; (reprint, Paris: Editions de Minuit, 1961, with afterword, "La victoire," by Jean-Paul Sartre); trans. John Calder as *The Question* (New York: Braziller, 1958).

Amar, Marianne, and Pierre Milza, *L'immigration en France au XX siècle* (Paris: Armand Colin, 1990).

Anderson, Benedict, *Imagined Communities: Reflections on the Origin and Spread of Nationalism* (London: Verso, 1983).

Angenot, Marc, *1889: Un état du discours social* (Québec: Editions du préambule, 1989).

Armes, Roy, *French Cinema since 1946,* vols. 1 and 2 (New York: A. S. Barnes, 1970).

Aron, Raymond, *L'opium des intellectuels* (Paris: Calmann-Lévy, 1955); trans. Terence Kilmartin as *The Opium of the Intellectuals* (London: Secker and Warburg, 1957).

Aron, Raymond, *France, Steadfast and Changing: The Fourth to the Fifth Republic* (Cambridge: Harvard University Press, 1960).

Aron, Raymond, *Les desillusions du progrès: Essai sur la dialectique de la modernité* (Paris: Calmann-Lévy, 1969).

Aron, Raymond, *Mémoires* (Paris: René Julliard, 1983).

Bair, Deirdre, *Simone de Beauvoir* (New York: Simon and Schuster, 1990).

Balandier, Georges, *Histoire des autres* (Paris: Stock, 1977).

Balibar, Etienne, *Les frontières de la démocratie* (Paris: La Découverte, 1992).

Balibar, Etienne, and Immanuel Wallerstein, *Race, Nation, Classe* (Paris: La Découverte, 1988); trans. Chris Turner as *Race, Nation, Class: Ambiguous Identities* (London: Verso, 1991).

Bardon, Jean-Pierre, *La révolution automobile* (Paris: Albin Michel, 1977).

Barthes, Roland, *Mythologies* (Paris: Seuil, 1957); trans. Annette Lavers as *Mythologies* (New York: Noonday, 1972); and trans. Richard Howard as *The Eiffel Tower* (New York: Noonday, 1979).

Barthes, Roland, "La voiture, projection de l'égo," *Réalités* 213 (1963).

Barthes, Roland, *Essais critiques* (Paris: Seuil, 1964); trans. Richard Howard as *Critical Essays* (Evanston: Northwestern University Press, 1972).

234

Barthes, Roland, "Réponses," *Tel quel* 47 (Autumn 1971).

Barthes, Roland, *Roland Barthes* (Paris: Seuil, 1975); trans. Richard Howard as *Roland Barthes* (New York: Hill and Wang, 1977).

Baudrillard, Jean, *Le système des objets* (Paris: Gallimard, 1968).

Baudrillard, Jean, *La société de consommation* (Paris: Editions Denoël, 1970).

Beauvoir, Simone de, *Les belles images* (Paris: Gallimard, 1966); trans. Patrick O'Brian as *Les belles images* (New York: Putnam, 1968).

Beauvoir, Simone de, *Tout compte fait* (Paris: Gallimard, 1972); trans. Patrick O'Brian as *All Said and Done* (New York: Putnam, 1974).

Beauvoir, Simone de, and Gisèle Hamini, *Djamila Boupacha* (Paris: Gallimard, 1962); trans. Peter Green as *Djamila Boupacha* (New York: Macmillan, 1962).

Becker, Jacques, "Enquette sur Hollywood," *Cahiers du cinéma* 54 (December 1955) pp. 71–76.

Benjamin, Walter, *Charles Baudelaire: A Lyric Poet in the Era of High Capitalism,* trans. Harry Zohn (London: New Left Books, 1973).

Bétourné, Olivier, and Aglaia Hartig, *Penser l'histoire de la Révolution* (Paris: La Découverte, 1989).

Birchall, Ian, "Imperialism and Class: The French War in Algeria," in *Europe and Its Others,* vol. 2, ed. Francis Barker, Peter Hulme, Margaret Iversen, and Diana Loxley. (Colchester: University of Essex, 1985).

Bloch, Ernst, "Nonsynchronism and the Obligation to Its Dialectics," trans. Mark Ritter, *New German Critique,* no. 11 (Spring 1977), pp. 22–38.

Boltanski, Luc, "Accidents d'automobile et lutte des classes," *Actes de la recherche en sciences sociales* no. 2 (March 1973), pp. 25–41.

Boltanski, Luc, "Les usages sociaux de l'automobile: Concurrence pour l'espace et accidents," *Actes de la recherche en sciences sociales* no. 2 (March 1975), pp. 25–49.

Boltanski, Luc, "America, America . . . le plan Marshall et l'importation du 'management,'" *Actes de la recherche en sciences sociales* no. 38 (1981), pp. 19–41.

Boltanski, Luc, *Les cadres: La formation d'un groupe sociale* (Paris: Editions de Minuit, 1982); trans. Arthur Goldhammer as *The Making of a Class: Cadres in French Society* (London: Cambridge University Press, 1987).

Borgé, Jacques, and Nicolas Viasnoff, *La 4CV* (Paris: Balland, 1976).

Borhan, Pierre, and Patrick Roegiers, eds., *René-Jacques* (Paris: Editions La Manufacture, 1991).

Borne, Dominique, *Petits bourgeois en révolte? Le mouvement poujade* (Paris: Flammarion, 1977).

Boudard, Alphonse, *La fermeture* (Paris: Editions Robert Laffont, 1986).

Braudel, Fernand, "Histoires et sciences sociales: La longue durée," *Annales ESC,* no. 17 (1958); reprinted in *Ecrits sur l'histoire* (Paris: Flammarion, 1969); trans. Sarah Matthews as *On History* (Chicago: University of Chicago Press, 1980).

Burke, Peter, *The French Historical Revolution: The Annales School 1929–89* (Stanford: Stanford University Press, 1990).

Castoriadis, Cornelius, "Le mouvement révolutionnaire sous le capitalisme moderne," *Socialisme ou barbarie* 31–33 (December 1960; April and December 1961); trans. David Ames Curtis as "Modern Capitalism and Revolution," in *Political and Social Writings,* Vol. 2 (Minneapolis: University of Minnesota Press, 1988).

Castoriadis, Cornelius, *La société française* (Paris: 10/18, 1979).

Céaux, Jean, Patrick Mazet, and Tuoi Ngo Hong, "Images et réalités d'un quartier populaire: Le cas de Belleville," *Espaces et sociétés,* no. 30–31 (July–December 1979), pp. 71–107.

Césaire, Aimé, *Discours sur le colonialisme* (Paris: Présence Africaine, 1955); trans. Joan Pinkham as *Discourse on Colonialism* (New York: Monthly Review Press, 1972).

Chevalier, Louis, *L'assassinat de Paris* (Paris: Calmann-Lévy, 1977); trans. David P. Jordan as *The Assassination of Paris* (Chicago: University of Chicago Press, 1994).

Closon, Francis-Louis, *Un homme nouveau: L'ingénieur économiste* (Paris: Presses universitaires françaises, 1961).

Communica International, *De la 4CV à la vidéo 1953–1983: Ces trente années qui ont changé notre vie* (Paris: Communica International, 1983).

Crozier, Michel, *Le phénomène bureaucratique* (Paris: Seuil, 1963); translated by the author as *The Bureaucratic Phenomenon* (Chicago: University of Chicago Press, 1964).

Crozier, Michel, "The Cultural Revolution: Notes on the Changes in the Intellectual Climate in France," *Daedalus* no. 2 (1964), pp. 514–542.

Daeninckx, Didier, *Meurtres pour mémoire* (Paris: Gallimard, 1981).

Daeninckx, Didier, *Le bourreau et son double* (Paris: Gallimard, 1986).

Dambre, Marc, *Roger Nimier: Hussard du demi-siècle* (Paris: Flammarion, 1989).

Debray, Régis, *A demain de Gaulle* (Paris: Gallimard, 1990).

de Grazia, Victoria, "Mass Culture and Sovereignty: The American Challenge to European Cinemas, 1920–1960," *Journal of Modern History* 61 (March 1989), pp. 53–87.

———

Deutschmann, David, ed., *Che Guevara and the Cuban Revolution: Writings and Speeches of Ernesto Che Guevara* (Sydney: Pathfinder, 1987).

Djébar, Assia, *Les enfants du nouveau monde* (Paris: René Julliard, 1962).

Dosse, François, *L'histoire en miettes: Des "Annales" à la "nouvelle histoire"* (Paris: La Découverte, 1987).

Dosse, François, *Histoire du structuralisme,* vol. 1 (Paris: La Découverte, 1991).

Dresch, Jean, "Le fait national algérien," *La pensée* (July 1956).

Droz, Bernard, and Evelyne Lever, *Histoire de la guerre d'Algérie 1954–1962* (Paris: Seuil, 1982).

Dubois, Jean, *Les cadres dans la société de consommation* (Paris: Editions du Cerf, 1969).

Dubois, Jean, *Les cadres, enjeu politique* (Paris: Seuil, 1971).

duBois, Page, *Torture and Truth* (New York: Routledge, 1991).

Dubois-Jallais, Denise, *La tzarine: Hélène Lazareff et l'aventure de 'ELLE'* (Paris: Editions Robert Laffont, 1984).

Duchen, Claire, "Occupation Housewife: The Domestic Ideal in 1950s France," *French Cultural Studies* 2 (1991), pp. 1–11.

Durand, Robert, *La lutte des travailleurs de chez Renault racontée par eux-mêmes, 1912–1944* (Paris: Editions sociales), 1971.

Duriez, Bruno, "De l'insalubrité comme fait politique," *Espaces et sociétés* no. 30–31 (July–December 1979), pp. 37–55.

Etcherelli, Claire, *Elise ou la vraie vie* (Paris: Editions Denoël, 1967); trans. by June Wilson and Walter Benn Michaels as *Elise or the Real Life* (New York: Morrow, 1969).

Eveno, Claude, and Pascale de Mezamat, eds., *Paris perdu: Quarante ans de boule-versements de la ville* (Paris: Editions Carré, 1991).

Evenson, Norma, *Paris: A Century of Change, 1878–1978* (New Haven: Yale University Press, 1979).

Fallaize, Elizabeth, *The Novels of Simone de Beauvoir* (London: Routledge, 1988).

Fanon, Frantz, *L'an V de la Révolution algérienne* (1959; reprinted as *Sociologie d'une révolution,* Paris: Maspero, 1968); trans. Haakon Chevalier as *A Dying Colonialism* (New York: Grove, 1965).

Fanon, Frantz, *Les damnés de la terre* (Paris: Maspero, 1961), trans. Constance Farrington as *The Wretched of the Earth* (New York: Grove, 1968).

Fanon, Frantz, *Pour la révolution africaine* (Paris: Maspero, 1964), trans. Haakon Chevalier as *Toward the African Revolution* (New York: Grove, 1967).

Field, Belden, "French Maoism," in *The 60s without Apology,* ed. Sohnya Sayers, Anders Stephanson, Stanley Aronowitz, and Fredric Jameson (Minneapolis: University of Minnesota Press, 1984).

Fitzgerald, Sean, "The Anti-Modern Rhetoric of Le Mouvement Poujade," *Review of Politics* 32 (April 1970), pp. 167–190.

Forty, Adrian, *Objects of Desire* (New York: Pantheon, 1986).

Foucault, Michel, *Les mots et les choses* (Paris: Gallimard, 1966); translated as *The Order of Things* (New York: Random House, 1970).

Fourastié, Jean, *Histoire du confort* (published 1950 as *Les arts ménagers*; reprint, Paris: PUF, 1973).

Fridenson, Patrick, *L'histoire des usines Renault* (Paris: Seuil, 1972).

Fridenson, Patrick, "La bataille de la 4CV," *L'histoire* 9 (February 1979).

Gabrysiak, Michel, *Cadres, qui êtes-vous?* (Paris: Editions Robert Laffont, 1968).

Gauron, André, *Histoire économique de la Vième république,* vol. 1 (Paris: Maspero, 1983).

Giroud, Françoise, *Françoise Giroud vous présente le tout-Paris* (Paris: Gallimard, 1952).

Giroud, Françoise, "Apprenez la politique," *Elle* (May 2, 1955).

Giroud, Françoise, *Si je mens . . .* (Paris: Stock, 1972); trans. Richard Seaver as *I Give You My Word* (Boston: Houghton Mifflin, 1974).

Giroud, Françoise, *Leçons particulières* (Paris: Livres de poche, 1990).

Godard, Jean-Luc, *Jean-Luc Godard par Jean-Luc Godard* (Paris: Editions Pierre Belfond, 1968).

Godard, Jean-Luc, *Introduction à une véritable histoire du cinéma,* vol. 1 (Paris: Editions Albatros, 1980).

Godelier, Maurice, "Functionalism, Structuralism, and Marxism," foreword to the English edition of *Rationalité et irrationalité en économie* (Paris: Maspero, 1966): *Rationality and Irrationality in Economics* (London: New Left Books, 1972), pp. vii–xlii.

Goldmann, Annie, *Cinéma et société moderne: Le cinéma de 1958 à 1968* (Paris: Editions Denoël, 1971).

Gorz, André, *Critique de la division du travail* (Paris: Seuil, 1973).

Grall, Xavier, *La génération du Djebel* (Paris: Editions du Cerf, 1962).

Grégoire, Ménie, "La presse feminine," *Esprit* (July–August 1959), pp. 17–34.

Guevara, Ernesto Che, *Le socialisme et l'homme* (1967; reprint, Paris: La Découverte, 1987).

Hamon, Hervé, and Patrick Rotman, *Les porteurs de valises: La résistance française à la guerre d'Algérie* (Paris: Seuil, 1982).

Hamon, Hervé, and Patrick Rotman, *Génération: Les années de rêve* (Paris: Seuil, 1987).

Harrison, Martin, "Government and Press in France during the Algerian War," *The American Political Science Review* 58, no. 2 (June 1964), pp. 273–285.

Harvey, David, *The Condition of Postmodernity: An Enquiry into the Origins of Cultural Change* (Oxford: Blackwell, 1989).

Herzlich, Guy, "Adieu Billancourt," *Le Monde* (March 29, 1992), p. 25.

Hoffmann, Stanley, *Le mouvement poujade* (Paris: Armand Colin, 1956).

Hoffmann, Stanley, R. R. Bowie, C. P. Kindleberger, L. Wylie, J. R. Pitts, J.-B. Duroselle, and F. Goguel, *In Search of France* (Cambridge: Harvard University Press, 1963).

Hollifield, James F., and George Ross, eds., *Searching for the New France* (New York: Routledge, 1991).

Jameson, Fredric, *Sartre: The Origins of a Style* (New York: Columbia, 1984).

Jameson, Fredric, *The Ideologies of Theory,* vol. 1 (Minneapolis: University of Minnesota Press, 1988).

Jameson, Fredric, *Postmodernism, or the Cultural Logic of Late Capitalism* (Durham: Duke University Press, 1991).

Judt, Tony, *Past Imperfect: French Intellectuals, 1944–1956* (Berkeley: University of California Press, 1992).

Kaplan, Alice, and Kristin Ross, "Everyday Life," *Yale French Studies* 73 (Fall 1987).

Keefe, Terry, *Simone de Beauvoir: A Study of Her Writings* (London: Harrap, 1983).

Kracauer, Siegfried, *Orpheus in Paris: Offenbach and the Paris of His Time,* trans. Gwenda David and Eric Mosbacher (New York: Knopf, 1938).

Kuisel, Richard, *Capitalism and the State in Modern France* (Cambridge: Cambridge University Press, 1981).

Kuisel, Richard, *Seducing the French: The Dilemma of Americanization* (Berkeley: University of California Press, 1993).

L'internationale Situationniste: 1958–69 (Paris: Champ Libre, 1970); ed. and trans. Ken Knabb as *Situationist International Anthology* (Berkeley: Bureau of Public Secrets, 1981).

Laubier, Claire, ed., *The Condition of Women in France: 1945 to the Present* (London: Routledge, 1990).

Lavadon, Pierre, *Nouvelle histoire de Paris: Histoire de l'urbanisme à Paris* (Paris: Hachette, 1975).

Leak, Andrew, "Phago-citations: Barthes, Perec, and the Transformations of Literature," *Review of Contemporary Fiction* 13, no. 1 (Spring 1993), pp. 57–75.

———

Lebrigand, Yvette, "Les archives du *salon des arts ménagers*," *Bulletin de l'institut d'histoire du temps présent* (December 1986), pp. 9–13.

Leenhardt, Jacques, *Lecture politique du roman* (Paris: Editions de Minuit, 1973).

Lefebvre, Henri, *Critique de la vie quotidienne,* 3 vols. (Paris: Arche, 1958–1981); Volume 1 trans. John Moore as *The Critique of Everyday Life* (London: Verso, 1991).

Lefebvre, Henri, *La somme et le reste* (Paris: La Nef de Paris, 2 volumes, 1959; reprint, Paris: Méridiens Klincksieck, 1989).

Lefebvre, Henri, *Position: Contre les technocrates* (Paris: Gonthier, 1967).

Lefebvre, Henri, *Le droit à la ville* (Paris: Anthropos, 1968).

Lefebvre, Henri, *La vie quotidienne dans le monde moderne* (Paris: Gallimard, 1968); trans. Sacha Rabinovitch as *Everyday Life in the Modern World* (New York: Harper, 1971).

Lefebvre, Henri, *Au delà du structuralisme* (Paris: Anthropos, 1971; reprinted as *L'idéologie structuraliste,* Paris: Anthropos, 1971).

Lefebvre, Henri, *La production de l'espace* (Paris: Anthropos, 1974); trans. Donald Nicholson-Smith as *The Production of Space* (Oxford: Basil Blackwell, 1991).

Lefebvre, Henri, *Le temps des méprises* (Paris: Stock, 1975).

Leulliette, Pierre, *St. Michel et le Dragon* (Paris: Editions de Minuit, 1961); translated as *St. Michael and the Dragon* (London: Heinemann, 1964).

Lévi-Strauss, Claude, *Tristes tropiques* (Paris: Librairie Plon, 1955); trans. John and Doreen Weightman as *Tristes tropiques* (New York: Penguin, 1973).

Lévi-Strauss, Claude, *La pensée sauvage* (Paris: Librairie Plon, 1962); trans. George Weidenfeld as *The Savage Mind* (Chicago: University of Chicago Press, 1966).

Lévi-Strauss, Claude, "Du bon usage du structuralisme," *Le monde* (January 13, 1968).

Lévi-Strauss, Claude, *De près et de loin* (Paris: Odile Jacob, 1988).

Lindon, Jerome, ed., *La gangrène* (Paris: Editions de Minuit, 1959); trans. Robert Silvers as *The Gangrene* (New York: Lyle Stuart, 1960).

Linhart, Robert, *L'établi* (Paris: Editions de Minuit, 1978), trans. Margaret Crosland as *The Assembly Line,* (Amherst: University of Massachusetts Press, 1981).

Lipietz, Alain, *Le capital et son espace* (Paris: La Découverte/Maspero, 1983).

Mallet, Serge, *La nouvelle classe ouvrière* (Paris: Seuil, 1969).

Mallet-Joris, Françoise, *Les signes et les prodiges* (Paris: Grasset, 1966).

Maran, Rita, *Torture: The Role of Ideology in the French-Algerian War* (New York: Praeger, 1989).

Marseille, Jacques, *Empire colonial et capitalisme français (années 1880–années 1950), histoire d'un divorce* (Paris: Albin Michel, 1984).

Marx, Karl, *Grundrisse: Foundations of the Critique of Political Economy,* trans. Martin Nicolaus (London: New Left Review, 1973).

Maspero, François, *Les passagers du Roissy-Express* (Paris: Seuil, 1990).

Mattleart, Armand, *La communication-monde* (Paris: La Découverte, 1992).

Maurienne [Jean-Louis Hurst], *Le déserteur* (Paris: Editions de Minuit, 1960; reprint, Paris: Editions Manya, 1991).

Mazon, B., "Fondations américaines et sciences sociales en France: 1920–1960," thesis, Ecole des hautes études en sciences sociales, 1985.

Memmi, Albert, *Portrait du colonisé précédé du portrait du colonisateur* (Paris: Editions Corréa, 1957); trans. Howard Greenfield as *The Colonizer and the Colonized* (New York: Orion Press, 1965).

Meynaud, Jean, *La technocratie, mythe ou réalité* (Paris: Payot, 1964).

Morin, Edgar, "La vie quotidienne et sa critique," *La NEF* 17 (May 1958), pp. 82–86.

Morin, Edgar, *L'esprit du temps* (Paris: Grasset, 1962).

Morin, Edgar, "Préface," to French translation of D. Riesman's *The Lonely Crowd: La foule solitaire* (Paris: B. Arthaud, 1964).

Morin, Edgar, *Commune en France: La métamorphose de Plodémet* (Paris: Fayard, 1967).

Mothé, Daniel, *Militant chez Renault* (Paris: Seuil, 1965).

Narboni, Jean, and Tom Milne, eds., *Godard on Godard* (New York: Viking, 1972).

Naville, Pierre, *L'état entrepreneur: Le cas de la Régie Renault* (Paris: Anthropos, 1971).

Ogden, Phillip, and Paul White, eds., *Migrants in Modern France* (London: Unwin Hyman, 1989).

Ollier, Claude, *La mise en scène* (Paris: Editions de Minuit, 1958).

Ory, Pascal, and Jean-Francois Sirinelli, *Les intellectuels en France: De l'affaire Dreyfus à nos jours* (Paris: Armand Colin, 1992).

Paxton, Robert, *Vichy France: Old Guard and New Order, 1940–1944* (New York: Columbia University Press, 1972).

Perec, Georges, *Les choses* (Paris: René Julliard, 1965); trans. David Bellos as *Things* (Boston: Godine, 1990).

Perec, Georges, "Georges Perec Owns Up: An Interview with Marcel Benabou and Bruno Marcenac," *The Review of Contemporary Fiction* 13, no. 1, (Spring 1993), pp. 17–20.

Poster, Mark, *Existential Marxism in Postwar France* (Princeton: Princeton University Press, 1975).

Poujade, Pierre, *J'ai choisi le combat* (Saint-Céré: Société générale des éditions et des publications, 1955).

Poulantzas, Nicolas, *Les classes sociales dans le capitalisme aujourd'hui* (Paris: Seuil, 1974); trans. David Fernbach as *Classes in Contemporary Capitalism* (London: New Left Books, 1975).

Rabinbach, Anson, *The Human Motor* (New York: Basic Books, 1990).

Rancière, Jacques, *La leçon d'Althusser* (Paris: Gallimard, 1974).

Régie nationale des usines Renault, *L'automobile de France,* text by Jules Romains, photos by René-Jacques (Paris: Régie nationale des usines Renault, 1951).

Rhodes, Anthony, *Louis Renault: A Biography* (New York: Harcourt, Brace and World, 1969).

Rifkin, Adrian, *Street Noises: Parisian Pleasure 1900–1940* (Manchester: Manchester University Press, 1993).

Rioux, Jean-Pierre, ed., *La France de la Quatrième République,* vols. 1 and 2 (Paris: Seuil, 1980–1983), trans. Godfrey Rogers as *The Fourth Republic 1944–1958* (London: Cambridge University Press, 1987).

Rioux, Jean-Pierre, ed., *La guerre d'Algérie et les Français* (Paris: Fayard, 1990).

Rioux, Jean-Pierre, and Jean-Francois Sirinelli, eds., *La guerre d'Algérie et les intellectuels français* (Paris: Editions Complexe, 1991).

Robbe-Grillet, Alain, *La jalousie* (Paris: Editions de Minuit, 1957); trans. Richard Howard as *Jealousy* (New York: Grove, 1959).

Robbe-Grillet, Alain, *Pour un nouveau roman* (Paris: Editions de Minuit, 1963); trans. Richard Howard as *For a New Novel* (New York: Grove, 1965).

Robbe-Grillet, Alain, *Le miroir qui revient* (Paris: Editions de Minuit, 1984).

Rochefort, Christiane, *Les petits enfants du siècle* (Paris: Grasset, 1961); translated as *Children of Heaven* (New York: David McKay, 1962).

Rochefort, Christiane, *Les stances à Sophie* (Paris: Grasset, 1963).

Roger, Philippe, *Roland Barthes, roman* (Paris: Editions Grasset & Fasquelle, 1986).

Ross, Kristin, *The Emergence of Social Space: Rimbaud and the Paris Commune* (Minneapolis: University of Minnesota Press, 1988).

Ross, Kristin, "Watching the Detectives," in *Postmodernism and the Reinvention of Modernity,* ed. Francis Barker, Peter Hulme, and Margaret Iversen, (Manchester: Manchester University Press, 1992).

247

Rostow, Walt Whitman, *The Stages of Economic Growth: A Non-communist Manifesto* (Cambridge: Cambridge University Press, 1960).

Sagan, Françoise, *Bonjour tristesse* (Paris: René Julliard, 1954); trans. Irene Ash as *Bonjour tristesse* (New York: Dell, 1955).

Sagan, Françoise, *Un certain sourire* (Paris: René Julliard, 1956).

Sagan, Françoise, *Aimez-vous Brahms . . .* (Paris: René Julliard, 1959); trans. Peter Wiles as *Aimez-vous Brahms . . .* (New York: Dutton, 1960).

Sagan, Françoise, "La jeune fille et la grandeur," *L'Express* (June 16, 1960); reprinted in English in Beauvoir and Hamini, *Djamila Boupacha*.

Sagan, Françoise, *Réponses* (Paris: Editions Jean-Jacques Pauvert, 1974); trans. David Macey as *Réponses: The Autobiography of Françoise Sagan* (Godalming, England: The Ram Publishing Company, 1979).

Sagan, Françoise, *Avec mon meilleur souvenir* (Paris: Gallimard, 1984); trans. Christine Donougher as *With Fondest Regards* (New York: Dutton, 1985).

Saint-Geours, Jean, *Vive la société de consommation* (Paris: Hachette, 1971).

Sartre, Jean-Paul, "Jean-Paul Sartre répond," *L'arc* 30 (1966).

Sayer, Derek, *The Violence of Abstraction* (Oxford: Blackwell, 1987).

Scardigli, Victor, *La consommation: Culture du quotidien,* (Paris: PUF, 1983).

Schivelbusch, Wolfgang, *The Railway Journey: Trains and Travel in the 19th Century* (New York: Urizen, 1979).

Servan-Schreiber, Jean-Jacques, *Lieutenant en Algérie* (Paris: René Julliard, 1957); trans. Ronald Matthews as *Lieutenant in Algeria* (New York: Knopf, 1957).

Servan-Schreiber, Jean-Jacques, *Le défi américain* (Paris: Denoël, 1967); trans. Ronald Steel as *The American Challenge* (New York: Atheneum, 1968).

Servan-Schreiber, Jean-Louis, *A mi-vie: L'entrée en quarantaine* (Paris: Stock, 1977).

Siritzky, Serge, and Françoise Roth, *Le roman de l'Express 1953–1978* (Paris: Atelier Marcel Jullian, 1979).

Stora, Benjamin, *La gangrène et l'oubli: La mémoire de la guerre d'Algérie* (Paris: La Découverte, 1991).

Sullerot, Evelyne, *La presse féminine* (Paris: Armand Colin, 1983).

Touraine, Alain, *L'évolution du travail ouvrier aux usines Renault* (Paris: Centre national de la recherche scientifique, 1955).

Touraine, Alain, "Travail, loisirs et société," *Esprit* (June 1959), pp. 979–999.

Touraine, Alain, *Sociologie de l'action* (Paris: Seuil, 1965).

Touraine, Alain, *La société post-industrielle* (Paris: Denoël, 1969).

Trinquier, Roger, *La guerre moderne* (Paris: Editions de la Table Ronde, 1961); trans. Daniel Lee as *Modern Warfare* (New York: Praeger, 1964).

Triolet, Elsa, *Roses à crédit,* vol. 1 of *L'age du nylon* (Paris: Gallimard, 1959).

Truffaut, François, "Feu James Dean," *Arts* (September 26, 1956), p. 4.

Turner, Dennis, "Made in the USA: Transformation of Genre in the Films of Jean-Luc Godard and François Truffaut," Ph.D. dissertation, University of Indiana, 1981.

Vallin, Jacques, and Jean-Claude Chesnais, "Les accidents de la route en France. Mortalité et morbidité depuis 1953," *Population* 3 (May–June 1975), pp. 443–478.

Vian, Boris, *Chansons et poèmes* (Paris: Editions Tchou, 1960).

Vidal-Naquet, Pierre, *La torture dans la république* (Paris: Editions de Minuit, 1972); originally published as *Torture: Cancer of Democracy* (Harmondsworth: Penguin, 1963).

Vittori, Jean-Pierre, *Nous, les appelés d'Algérie* (Paris: Stock, 1977).

Weil, Simone, *La condition ouvrière* (Paris: Gallimard, 1951).

White, Paul E., "Immigrants, Immigrant Areas and Immigrant Communities in Postwar Paris," in *Migrants in Modern France,* ed. Philip Ogden and Paul E. White (London: Unwin Hyman, 1989), pp. 195–212.

Winock, Michel, *Chronique des années soixante* (Paris: Seuil, 1987).

Wolf, Eric, *Peasant Wars of the Twentieth Century* (New York: Harper and Row, 1969).

Zola, Emile, *Au bonheur des dames* (Paris: Gallimard, 1980); translated anonymously, with critical introduction by Kristin Ross, as *The Ladies' Paradise* (Berkeley: University of California Press, 1992).

SELECTED FILMOGRAPHY

Carné, Marcel, *Les tricheurs* (1958).

Chabrol, Claude, *Le beau Serge* (1958).

Chabrol, Claude, *Les cousins* (1958).

Chabrol, Claude, *Les bonnes femmes* (1960).

Demy, Jacques, *Lola* (1960).

Demy, Jacques, *Les parapluies de Cherbourg* (1964).

Dhery, Robert, *La belle américaine* (1961).

Godard, Jean-Luc, *A bout de souffle* (1959).

Godard, Jean-Luc, *Une femme est une femme* (1961).

Godard, Jean-Luc, *Le mépris* (1963).

Godard, Jean-Luc, *Bande à part* (1964).

Godard, Jean-Luc, *Alphaville* (1965).

Godard, Jean-Luc, *Pierrot le fou* (1965).

Godard, Jean-Luc, *Masculin/féminin* (1966).

Godard, Jean-Luc, *Made in U.S.A.* (1966).

Godard, Jean-Luc, *Weekend* (1967).

Lelouch, Claude, *Un homme et une femme* (1966).

Malle, Louis, *Ascenseur pour l'échafaud* (1957).

Marker, Chris, *Le joli mai* (1962).

Pontecorvo, Gilles, *La bataille d'Alger* (1967).

Preminger, Otto, *Bonjour tristesse* (1957).

Resnais, Alain, *Muriel* (1963).

Risi, Dino, *Il sorpasso* (1962).

Rouche, Jean, and Edgar Morin, *Chronique d'un été* (1961).

Rozier, Jacques, *Adieu Philippine* (1962).

Tati, Jacques, *Jour de fête* (1949).

Tati, Jacques, *Les vacances de M. Hulot* (1953).

Tati, Jacques, *Mon oncle* (1958).

Tati, Jacques, *Playtime* (1967).

Truffaut, François, *Jules et Jim* (1961).

Varda, Agnès, *Cléo de 5 à 7* (1962).

Index